PENGUIN BOOKS

THE PENGUIN BOOK OF CURIOUS AND INTERESTING MATHEMATICS

David Wells was born in 1940. He had the rare distinction of being a Cambridge scholar in mathematics and failing his degree. He subsequently trained as a teacher and, after working on computers and teaching machines, taught mathematics and science in a primary school and mathematics in secondary schools. He is still involved with education through writing and research.

While at university he became British under-21 chess champion, and in the mid-seventies was a game inventor, devising 'Guerilla' and 'Checkpoint Danger', a puzzle composer, and the puzzle editor of *Games & Puzzles* magazine. From 1981 to 1983 he published *The Problem Solver*, a magazine of mathematical problems for secondary pupils. He has published several books of problems and popular mathematics, including *Can You Solve These?* and *Hidden Connections, Double Meanings*, as well as *Russia and England, and the Transformation of European Culture*. He is also the author of *The Penguin Book of Curious and Interesting Puzzles*, *The Penguin Book of Curious and Interesting Numbers*, *The Penguin Dictionary of Curious and Interesting Geometry* and *You Are a Mathematician*.

David Wells

The Penguin Book of Curious and Interesting Mathematics

PENGUIN BOOKS

PENGUIN BOOKS

Published by the Penguin Group
Penguin Books Ltd, 27 Wrights Lane, London W8 5TZ, England
Penguin Putnam Inc., 375 Hudson Street, New York, New York 10014, USA
Penguin Books Australia Ltd, Ringwood, Victoria, Australia
Penguin Books Canada Ltd, 10 Alcorn Avenue, Toronto, Ontario, Canada M4V 3B2
Penguin Books (NZ) Ltd, 182–190 Wairau Road, Auckland 10, New Zealand

Penguin Books Ltd, Registered Offices: Harmondsworth, Middlesex, England

First published 1997
10 9 8 7 6 5 4 3

Set in 10/12pt Monotype Sabon by
Rowland Phototypesetting Limited
Bury St Edmunds, Suffolk
Printed in England by
Clays Ltd, St Ives plc

Contents

Introduction

Welcome to this collection of anecdotes and stories, snatches of history, human foibles and amazing facts, not forgetting the occasional tragic deathbed scene, odd formula and literary curiosity.

Mathematics is a wonder. Its historical roots can be traced far back into the origins of every civilization. Its psychological roots are embedded in the most ordinary activities of everyday life. It is more than an art, more than a science, more than a game. It is a recreation, but also potentially, an obsession: the first and purest science, but also a mysteriously and incomprehensibly powerful tool for understanding the real world.

Surprisingly, in view of its immense history, little or nothing is known about what makes a mathematician. Why should this apparently abstract and dry study prove entrancing to so many different individuals?

Mathematicians have come from every walk of life, and every background. They vary from the academic who fits every stereotype of the mad professor, to the man of the world for whom mathematics is a relaxation. Amateur mathematicians have belonged to every other profession, while professionals have espoused every political position, and every philosophical perspective. They have been male and female, but many fewer of the latter – why? They have produced wonderful work of lasting importance before dying tragically young, and continued productively into old age, blind but not beaten. In all their extraordinary variety, they are the subject of this book.

This is not a history, nor are entries arranged in any particular order, least of all chronological. I have merely dipped into a vast cornucopia and pulled out those plums that most appealed to me and that I hoped would be enjoyed by readers. I have tried to be catholic in my tastes, but there will be inevitable biases. I offer my apologies in advance to readers who find that their favourite story is *not* included.

Mathematics is a profoundly human activity. If this collection impresses that thought on readers, and provides some serious insights into the lives of the great, and some not so great, mathematicians, while also providing light-hearted entertainment, I shall be satisfied.

Acknowledgements

The author and publisher are grateful to the following for permission to reproduce copyright material:

'Cardano wheedles the secret of the cubic out of Tartaglia', reprinted from J. Fauvel and J. Gray (eds.), *The History of Mathematics: A Reader*, 1987, pp. 254–5, by permission of the Open University Press. 'The most prolific mathematician of all time', 'Euler becomes blind' and 'Euler as a marine engineer', reprinted from C. Truesdell, 'Leonard Euler, supreme geometer (1701–1783)', in H. E. Pagliaro (ed.), *Irrationalism in the Eighteenth Century*, pp. 51, 84–5, by permission of American Society for Eighteenth-Century Studies. 'Alonzo Church', reprinted from G.-C. Rota, 'Fine Hall in its golden era', in *A Century of Mathematics in America*, 1989, pp. 224–5, by permission of the American Mathematical Society. 'An extraordinary mind', reprinted from G.-C. Rota, 'The Lost Café', in *Contention*, vol. 2, 2, Winter 1993, pp. 42–3, 51, by permission of G.-C. Rota. 'The USSR against G. H. Hardy', reprinted from A. M. Vershik, '"Nature" and Soviet censorship', in *London Mathematical Society Newsletter*, 218, July 1994, p. 12, by permission of the London Mathematical Society and Professor A. M. Vershik. 'John Horton Conway', reprinted from D. J. Albers and G. L. Alexanderson (eds.), *Mathematical People, Profiles and Interviews*, © 1984, used with permission of Contemporary Books, Inc., Chicago. 'Julia Robinson and Hilbert's Tenth Problem', 'Boy from the word go' and 'Grace Chisholm Young', reprinted from D. J. Albers, G. L. Alexanderson and C. Reid (eds.), *More Mathematical People*, 1990, pp. 222, 224, 266–7, 299, by permission of Harcourt, Brace, Jovanovich. 'The death of John von Neumann', 'A prodigious childhood' and 'Wiener's and von Neumann's methods of working' reprinted from S. J. Heims, *John von Neumann and Norbert Wiener: From Mathematics to the Technologies of Life and Death*, 1980, pp. 5–9, 123–7, 369–71, by permission of MIT Press. 'Mathematical pictures in physics', reprinted from R. P. Feynman, *The Development of the Space-Time View of Quantum Mechanics*, Les prix Nobel, 1966, by permission of The Nobel Foundation, © 1960.

Saunderson's pin-board, reproduced by permission of Cambridge University Press, from M. J. Morgan, *Molyneaux's Question*, Cambridge University Press, 1977, p. 44. The dining philosophers problem, reproduced by permission of Addison-Wesley Inc. from D. Harel, *Algorithmics: The Spirit of Computing*, Addison-Wesley, 1987, p. 287. Gauss's brain, reproduced by permission of Stephen Jay Gould from *The Mismeasure of Man*, W. W. Norton, 1981, p. 93. Leonardo's parachute, reproduced by permission of Professor G. A. Tokaty, from G. A. Tokaty, *A History and Philosophy of Fluidmechanics*, G. T. Foulis and Co., 1971, p. 47, fig. 22. The pons asinorum, reproduced by permission of Everyman's Library, from *The Elements of Euclid*, Everyman's Library, no. 891, J. M. Dent, 1948, p. 11. The Lanchester Model, reproduced by permission of Springer Verlag, New York, from M. Braun, *Differential Equations and their Applications*, Springer, New York, 1978, p. 298, fig. 3. Hero's steam engine, and a jet of steam supporting a sphere, reproduced by permission of Little, Brown & Co. (UK), from M. B. Hall (ed.), *The Pneumatics of Hero of Alexandria*, Macdonald, 1971, pp. 68, 72.

Every effort has been made to contact copyright holders. The publishers are willing to correct any omissions in future editions.

Curious and Interesting Mathematics

The mathematician, the physicist and the engineer

Mathematicians have always considered themselves superior to physicists and, of course, engineers. A mathematician gave this advice to his students: 'If you understand a theorem and you can prove it, publish it in a mathematics journal. If you understand it but can't prove it, submit it to a physics journal. If you can neither understand nor prove it, send it to a journal of engineering.'

Freeman Dyson, a brilliant mathematician as well as a world-famous physicist, disagreed. He wrote, only partly tongue-in-cheek – modern physics is both highly imaginative and obscure – that, 'Most of the papers which are submitted to the *Physical Review* are rejected, not because it is impossible to understand them, but because it is possible. Those that are impossible to understand are usually published.'

On the shores of the unknown

The Greek philosopher Aristippus was shipwrecked upon a strange shore, when he noticed a geometrical drawing in the sand. 'Be of good hope,' he said to his companions, 'for I see the footprints of men.'

Jules Verne imagined using the same idea to send a message to men in the moon. According to his novel *From the Earth to the Moon*,

'. . . a few days ago a German geometrician proposed to send a scientific expedition to the steppes of Siberia. There, on those vast plains, they were to describe enormous geometric forms, drawn in characters of reflecting luminosity, among which was the proposition regarding 'the square of the hypotenuse', commonly called the 'Ass's Bridge' by the French. 'Every intelligent being,' said the geometrician, 'must understand the scientific meaning of the figure. The Selenites, do they exist, will respond by a similar figure; and, a communication being at once established, it will be easy to form an alphabet which shall enable us to converse with the inhabitants of the moon.'

The same principle is used in modern attempts to communicate with the possible inhabitants of planets orbiting distant stars: the digits of π, or the details of a geometrical figure, are coded in binary digits, on the assumption that intelligent life anywhere, which is sufficiently advanced to receive radio signals, will be able to decode the message.

The physicist Eddington had a different angle. In his rather mystical philosophy he supposed that if only we could understand the workings of our own minds sufficiently well, then we would be able to understand the whole universe. Hence, he wrote:

'We have found a strange footprint on the shores of the unknown. We have devised profound theories, one after another, to account for its origin. At last, we have succeeded in reconstructing the creature that made the footprint. And Lo! it is our own.'

A slick limerick trick

A mathematician called Klein,
Thought the Moebius band was divine,
He said, 'If you glue,
The edges of two,
You'll get a weird bottle like mine!'

Leo Moser

A tricky choice

'E. Kummer, the German algebraist, was rather poor at arithmetic. Whenever he had occasion to do simple arithmetic in class he would get his students to help him. Once he had occasion to find 7 × 9. "7 × 9", he began, "7 × 9 is er – ah – ah – 7 × 9 is . . ." 61, a student suggested. Kummer wrote 61 on the board. "Sir," said another student, "it should be 69." "Come, come gentlemen," said Kummer, "it can't be both – it must be one or the other." '

Kummer had a famous predecessor, if *Spence's Anecdotes* is to be believed, in which Pope is quoted as saying that, 'Sir Isaac Newton, though so deep in Algebra and Fluxions, could not readily make up a common account; and, when he was Master of the Mint, used to get somebody to make up his accounts for him.'

Hilbert's memory

'Hasse once expressed to Mrs Hilbert a desire to speak personally with the great mathematician. Mrs Hilbert accordingly invited Hasse to tea one afternoon and then left him in the garden with her husband. Hasse soon launched into a discussion of class-field theory, a subject that had been created by Hilbert and that was of great interest to Hasse at the time. Hasse had written a report in the theory that had continued the earlier work done by Hilbert, and Hasse started to tell Hilbert of his added contributions. But Hilbert repeatedly interrupted Hasse, insisting that Hasse first explain the basic concepts and foundations of class-field theory. Hasse did so, and Hilbert grew enthusiastic and finally exclaimed, "All this is extremely beautiful; who created it?" And Hasse had to tell the astonished Hilbert that it was Hilbert himself who had created the beautiful theory.'

Conway's Last Theorem

The last theorem in John Conway's book, *On Numbers and Games*, is:

'Theorem 100. This is the last theorem in this book.
(The proof is obvious.)'

Joe Miller's Jest

'A famous Teacher of *Arithmetick*, who had long been married without being able to get his Wife with Child: One said to her, Madam, your Husband is an excellent *Arithmetician*. Yes, replied she, only he can't *multiply*.'

What is mathematics made of?

'A mathematician, like a painter or poet, is a maker of patterns. If his patterns are more permanent than theirs, it is because they are made with ideas.'

G. H. Hardy

Mersenne much earlier had argued that mathematics is 'a science of the imagination of pure intellect'. He wrote that certainty is possible in mathematics since it deals with quantities, it is 'a science of the imagination or of pure intellect; like metaphysics, which is not concerned with any other subject than what is possible in the absolute,' while Diderot said that the mathematician was like a gambler; they both played games with abstract rules that they themselves had created.

Why ten?

'Why do all men, barbarians and Greeks alike, count up to 10 and not up to any other number, saying for example, 2, 3, 4, 5 and then repeating them, "one-five", "two-five", just as they say eleven, twelve? Or why do they not stop at some point beyond ten and repeat from there? For every number is made up of one, two, &c., combined with a preceding number, and thus a different number is formed; but the counting always proceeds in fixed sets of ten. For it is clearly not the result of chance that all men invariably count in tens; and that which is invariable and universal is not the result of chance, but is in the nature of things. Is it because ten is a perfect number? For it combines every kind of number, odd and even, square and cube, length and surface, prime and composite. Or is it because ten is the original number, since one, two, three, and four together make ten? Or is it because the bodies which move in the heavens are nine in number? Or is it because in ten proportions four cubic numbers result, from which numbers the Pythagoreans declare that the whole universe is constituted? Or is it because all men have ten fingers, and so, as though possessing counters that indicate the numbers proper to man, they count all other things by this quantity? One race among the Thracians alone of all men count in fours, because their memory, like that of children, cannot extend farther and they do not use a large number of anything.'

Aristotle

The largest known prime number

At present, the largest known prime is $2^{756839}-1$, a number of 227,832 digits. In 1978, the largest prime was slightly smaller.

November 14, 1978

'Press Release

Two 18-year-old youths, Laura Nickel and Curt Noll, have calculated the largest known prime number, 2 to the 21,701st power minus one, using a terminal at California State University, Hayward and hooking into the CYBER 174 in the Los Angeles area. Totaling 6,533 digits, the number was proved to be prime last October 30 after three years of diligent study and work interrupted at times by official and personal problems.

Noll, a senior at Hayward High School, is also a freshman at Cal State enrolled in special classes. Nickel, a former Hayward High student, currently resides in Oakland.

Nickel and Noll became interested in computers about three years ago at Hayward High where they learned the basics of computer programming on the school's small computers. They became intrigued with Mersenne primes after seeing a poster which named the country's ten top young scientists, one chosen because he had computed large powers of two. They determined then to duplicate his effort.

As their enthusiasm for the project grew, they contacted Professor Arthur Simon, chairman of the Cal State math department, and Dr Dan Jurca, math lecturer who provided them with information and resource material about primes. Dr Jurca, who gave them access to the University's computer, said that their project was doomed. The tenacious teens vowed "to prove him wrong".

According to Nickel and Noll the period from June 1977 to August 1978 was the time of their "darkest hours". Over 1,900 hours were spent on two versions of the program which were not succeeding, and Nickel's parents did not approve of her spending so much time on the project. They were also hindered by a lack of certain facts which could only be obtained from a library in Los Angeles. Noll convinced his parents to take a vacation there, and he managed to secretly photocopy the needed material.

The long period of frustration ended on September 12 when the fifth version of the program succeeded.

This program started its search at 2 p.m. October 11, and at 9 p.m. on October 30 after 44 tests and approximately 440 hours, two to the 21,701st power minus one was found to be prime. Tests could be run only when the computer was not being used by the University for business or instruction.

Presently Nickel and Noll are continuing to search for larger primes . . .

Noll is employed as computer lab assistant at Cal State Hayward and is involved in several administration and student programming projects. Next year he will begin work on a double major in physics and computer science.

Nickel intends to attend college part time and major in philosophy and work part time in the area of computer science.'

Chrystal clear

G. Chrystal (1851–1911) was a well-known teacher and author of textbooks.

' "In the middle of some brilliant reasoning," recalls Barrie, "Chrystal would stop to add 4, 7, and 11. Addition of this kind was the only thing he could not do, and he looked to the class for help – '20,' they shouted, '24,' '17,' while he thought it over. These appeals to their intelligence made them beam."

The benches in Professor Chrystal's classroom rose in steps above the front floor. One day a student at the end of bench ten dropped a marble while the professor, with his back to the class, was working at the blackboard. The marble rolled downward toward the professor, very audibly falling down each of the ten steps on its way to the floor. Professor Chrystal never turned his head, but when the marble reached the floor, he said, still without turning his head, "Will the student at the end of bench ten, who dropped that marble, stand up?" While continuing his work at the blackboard, he had kept count of the falls of the marble from step to step.'

How to be a Good Lecturer

' "Well hello and welcome to the first lecture of the course er look I said hello look I'd like to start now will you shut up SHUT UP PLEASE oh thank you I don't mind you talking if you do so quietly I didn't ask to do this course you know I wanted to do algebra I told them I didn't know any analysis . . .

"... Now this course is all about complex numbers and I've got a list of recommended books here er well no in fact I seem to have left it behind never mind they're all out of print anyway now let me write up a definition where's the chalk gone ah here it is (SNAP) ah let me take another piece (THUD) not very big these platforms are they I keep falling off them . . .

"... Now definition 1.1 is ah um of course I haven't said what this section's called yet oh it doesn't seem to have a name anyway it's all about convergence of power series you did something like it in real analysis didn't you don't you remember well he should have done it in his lectures I don't have time to go into it now . . .

"... Now definition 1.1 (scribble scribble) can you read that at the back no oh well sit further forward then you can read it at the front ah come to think of it I can't read it either perhaps if I turn on this light ah no not that one another one oh well the cord was a bit frayed I suppose well look that symbol is a capital sigma yes what's the problem yes well green seems to be the only colour they have left in the box probably because nobody in his right mind uses it so they leave it for me . . .

"... Well look perhaps if I explain it in words it's all in the textbooks anyway I can't help it if they're missing from the library people eat them or something well now I'll draw a diagram you don't have to copy this exactly because it's slightly wrong anyway this is diagram 2 good question I think I forgot to draw diagram 1 anyway as I say it doesn't help much phew let me take my jacket off a bit (rip) oh well I sewed that button on myself you can tell can't you . . .

"... Now let me digress a minute about the history of the subject here it was discovered by Cauchy or do I mean Gauss one of those people and he sent a copy of his paper to someone else who well anyway it's very important and has a lot of applications such as er such as well anyway you will see applications in your other courses I expect of course they don't use the same notation but then they don't have the same ideas of rigour as we do and now let's write down the first result lemma 1.2 ...

"... Lemma 1.2 oh I haven't actually defined radius of convergence yet have I still let me write it up and we can decide what it means later well I still seem to have a few minutes left so I'd better start the proof let n be this and r be this and v be that and n be that no on second thoughts I'm already using n now so I'll call it nu pardon no it's a nu a greek letter you must have seen it before you know greek letters alpha etcetera no this one is nu all right call it v if you like but we're already using v still it won't cause confusion ...

"... Now multiply this out and obviously what we get is er clearly um oh that can't be right what have I done wrong here can you see the mistake maybe I lost a minus sign somewhere search me oh dear it's time to finish isn't it well give me just 5 more minutes and I'll finish this off and oh maybe I should do this bit again more carefully next time ah that should have been a nu maybe no it should be a v oh it's an r is it oh well look I'll finish this next time I'm sure I've got most of the details right it's really very elementary after all I haven't done anything nontrivial yet ..." '

Jonathan Partington

Louis Pósa, by Paul Erdös

'I will talk about Pósa who is now 22 years old and the author of about 8 papers. I met him before he was 12 years old. When I returned from the United States in the summer of 1959 I was told about a little boy whose mother was a mathematician and who knew quite a bit about high school mathematics. I was very interested and the next day I had lunch with him. While Pósa was eating his soup I asked him the following question: Prove that if you have $n+1$ positive integers less than or equal to $2n$, some pair of them are relatively prime. [That is, have no common factor, other than one.] It is quite easy to see that the claim is not true of just n such numbers because no two of the n even numbers up to $2n$ are relatively prime. Actually I discovered this simple result some years ago but it took me about ten minutes to find the really simple proof. Pósa sat there eating his soup, and then after a half a minute or so he said, "If you have $n+1$ positive integers less than or equal to $2n$, some two of them will have to be consecutive and thus relatively prime." Needless to say, I was very much impressed, and I venture to class this on the same level as Gauss' summation of the positive integers up to 100 when he was just 7 years old.'

Paul Erdös

The flies on Fliess

'One of the most extraordinary and absurd episodes in the history of numerological pseudoscience concerns the work of a Berlin surgeon named Wilhelm Fliess. Fliess was obsessed by the numbers 23 and 28. He convinced himself and others that behind all living phenomena and perhaps inorganic nature as well there are two fundamental cycles: a male cycle of 23 days and a female cycle of 28 days. By working with multiples of those two numbers – sometimes adding, sometimes subtracting – he was able to impose his number patterns on virtually everything. The work made a considerable stir in Germany during the early years of this century. Several disciples took up the system, elaborating and modifying it in books, pamphlets, and articles. In recent years the movement has taken root in the United States.

Although Fliess's numerology is of interest to recreational mathematicians and students of pathological science, it would probably be unremembered today were it not for one almost unbelievable fact: For a decade Fliess was Sigmund Freud's best friend and confidant.

Fliess's theory of cycles was at first regarded by Freud as a major breakthrough in biology. He sent Fliess information on 23- and 28-day periods in his own life and the lives of those in his family, and he viewed the ups and downs of his health as fluctuations of the two periods. He believed a distinction he had found between neurasthenia and anxiety neurosis could be explained by the two cycles. In 1898 he severed editorial connections with a journal because it refused to retract a harsh review of one of Fliess's books.

There was a time when Freud suspected that sexual pleasure was a release of 23-cycle energy and sexual unpleasure a release of 28-cycle energy. For years he expected to die at the age of 51 because it was the sum of 23 and 28, and Fliess had told him this would be his most critical year. "Fifty-one is the age which seems to be a particularly dangerous one to men," Freud wrote in his book on dreams. "I have known colleagues who have died suddenly at that age, and amongst them one who, after long delays, had been appointed to a professorship only a few days before his death."

Freud admitted on many occasions that he was hopelessly deficient in all mathematical abilities. Fliess understood elementary arithmetic, but little more. He did not realize that if any two positive integers that have no common divisor are substituted for 23 and 28 in his basic formula, it is possible to express *any positive integer whatever*.'

What is a proof?

'Some people believe that a theorem is proved when a logically correct proof is given; but some people believe that a theorem is proved only when the student sees why it is inevitably true. The author tends to belong to this second school of thought.'

Richard Hamming

From Cicero to typewriters

'It is impossible to calculate accurately events that are determined by chance.'

Thucydides *c.* 400 BC

Thucydides was wrong, though to be fair, he was an historian. Five hundred years later, Cicero, considering questions *Of Divination*, wrote, in a truly remarkable passage,

'Four dice are cast and a Venus throw results – that is chance; but do you think it would be chance, too, if in one hundred casts you made one hundred Venus throws? It is possible for paints flung at random on a canvas to form the outlines of a face; but do you imagine that an accidental scattering of pigments could produce the beautiful portrait of Venus of Cos? Suppose that a hog should form the letter "A" on the ground with its snout; is that a reason for believing that it could write out Ennius's poem *The Andromache*?

Carneades used to have a story that once in the Chian quarries when a stone was split open there appeared the head of the infant god Pan; I grant that the figure may have borne some resemblance to the god, but assuredly the resemblance was not such that you could ascribe the work to a Scopas. For it is undeniably true that no perfect imitation of a thing was ever made by chance.'

Cicero, of course, had no conception of probabilities as numbers, or as ratios – such ideas had to wait for Cardan, Pascal and Fermat. Yet of course he makes judgements, as we all do in everyday life, and reaches a conclusion that has a bizarre reflection in the idea of monkeys on typewriters.

Bertrand Russell started this particular hare:

'There is a special department of hell for students of probability. In this department there are many typewriters and many monkeys. Every time that a monkey walks on a typewriter, it types by chance one of Shakespeare's sonnets.'

Eddington added his sixpennyworth by composing this limerick:

There once was a brainy baboon
Who always breathed down a bassoon,
For he said, 'It appears
That in billions of years
I shall certainly hit on a tune.'

Aha! Here we have another conception that Cicero lacked, the idea of vast *but calculable* eras of time. This prompted the statement at a meeting many years ago of the British Association for the Advancement of Science, that 'If six monkeys were set before six typewriters it would be a long time before they produced by mere chance all the written books in the British museum; but it would not be an infinitely long time.'

The statement about the six stenographic monkeys inspired these verses, which appeared in the *Manchester Guardian*:

Life is brief, but art is longer
So the sages say in sooth –
Nothing could be worse or wronger
Than to doubt this ancient truth.
Endless volumes, larger, fatter
Prove man's intellectual climb,
But in essence it's a matter
Just of having lots of time.

Give me half a dozen monkeys,
Set them to the lettered keys,
And instruct these simian flunkies
Just to hit them as they please.
Lo! The anthropoid plebians
Toiling at their careless plan
Would in course of countless aeons
Duplicate the lore of man.

Thank you, thank you, men of science!
Thank you, thank you, British Ass!
I for long have placed reliance
On the tidbits that you pass.
And this season's nicest chunk is
Just to sit and think of those
Six imperishable monkeys
Typing in eternal rows!

Lucio

How long would it take for the monkeys to type something less extensive, say, the phrase, 'Dear Sir'? It depends on the number of keys on the typewriter and the speed of the monkey. Published estimates, for one monkey, have ranged from 1½ million to 1,365 million years, which suggests a degree of randomness in the calculations.

The idea soon appealed to science fiction writers. Raymond

F. Jones exploited the theme in 'Fifty Million Monkeys', which appeared in 1943 in *Astounding Science Fiction* magazine. J. J. Coupling wrote a skit on the *Mullabfuhrwortsmaschine*, a propaganda-generating device which could be adjusted to associate any groups or qualities to any desired degree.

Here is a shorter fantasy by Bruce Elliott, reprinted from Clifton Fadiman's *Fantasia Mathematica*:

FEARSOME FABLE

After they put the fifteen apes in front of the typewriters there was a long wait. The animals sat and looked at the machines, at the paper on the rollers. There was a long pause, then each ape, one after the other, leaned forward and typed a single, different word.

The experimenter waited a long, long time. But after the one flurry of activity, nothing happened. Finally, seeing that the apes had no intention of continuing, he went toward the typewriters.

The first ape had typed NOW; the second had typed IS; the third one, THE; the fourth, TIME; the fifth, FOR; the sixth ape, ALL; the seventh, GOOD; the eighth one, PARTIES; the ninth, TO; the tenth, COME; the eleventh, TO; the twelfth ape, THE; the thirteenth, AID; the fourteenth, OF; and the last ape had typed MAN.

To conclude this fantasia, it is true that *if* the mysterious number π is truly random, then every finite sequence of numbers will appear in it somewhere, (indeed, an infinite number of times). However, no one has yet discovered a single one of Shakespeare's sonnets, suitably coded, in the first few million digits of π, though, on the other hand, this could be because no one has yet looked for them . . .

Barrow's humour

Isaac Barrow (1630–1677) was a troublesome child, whose 'father was heard to pray that should God decide to take one of his children he could best spare Isaac'. He grew up to be the first Lucasian Professor of Mathematics at Cambridge, which post he gave up in 1670 to his pupil Isaac Newton.

'There was no love lost between Barrow and King Charles II's favorite, the Earl of Rochester, who had called the clergyman "a musty old piece of divinity". One day at court, where Barrow was serving as the king's chaplain, he encountered the earl, who bowed low and said sarcastically, "Doctor, I am yours to my shoe-tie."

"My lord," returned Barrow, "I am yours to the ground."

"Doctor, I am yours to the center."

"My lord, I am yours to the antipodes."

"Doctor, I am yours to the lowest pit of hell."

"And there, my lord, I leave you," said Barrow, turning smartly on his heel.'

Sylvester and *The Laws of Verse*

'To illustrate a theory of versification contained in his book *The Laws of Verse*, Sylvester prepared a poem of 400 lines, all rhyming with the name Rosalĭnd or Rosalīnd; and it was announced that the professor would read the poem on a specified evening at a specified hour at the Peabody Institute. At the time appointed there was a large turn-out of ladies and gentlemen. Prof. Sylvester, as usual, had a number of footnotes appended to his production; and he announced that in order to save interruption in reading the poem itself, he would first read the footnotes. The reading of the footnotes suggested various digressions to his imagination; an hour had passed, still no poem; an hour and a half passed and the striking of the clock or the unrest of his audience reminded him of the promised poem. He was astonished to find how time had passed, excused all who had engagements, and proceeded to read the Rosalind poem.'

The peripatetic Paul Erdös

Paul Erdös (1913–1996) is the world's only peripatetic mathematician, of 'no fixed abode', travelling from one mathematics department to another, owning the contents of two small suitcases, and spending his life solving problems and writing papers with hundreds of other more orthodox mathematicians.

Erdös was as careless of his personal state – he owned neither a chequebook nor a credit card – as he was generous to others.

In 1984 he won the $50,000 Wolf Prize, of which he kept $750 and gave the rest away.

An Erdös number was given to mathematicians who had co-authored a paper with someone who has co-authored a paper with someone . . . who had co-authored a paper with Erdös himself. If you once actually wrote a paper with Erdös, then your Erdös number is 1, and so on.

'Paul Erdös is the consummate problem solver: his hallmark is the succinct and clever argument [. . .] He loves areas of mathematics which do not require an excessive amount of technical knowledge but give scope for ingenuity and surprise. The mathematics of Paul Erdös is the mathematics of beauty and insight.

One of the most attractive ways in which Paul Erdös has influenced mathematics is through a host of stimulating problems and conjectures, to many of which he has attached money prizes, in accordance with their notoriety. He often says that he could not pay up if all his problems were solved at once, but neither could the strongest bank if all its customers withdrew their money at the same time. And the latter is far more likely.

Ever since he was greeted at the station in Cambridge by Davenport and Rado in October 1934, on his way from Hungary to Manchester to work with Mordell, Paul Erdös has travelled the world constantly: he is the archetypal peripatetic mathematician. As the "Professor of the Universe", he travels light, saying that "private property is a nuisance"; wherever he arrives his "brain is open" for new problems and ideas: "another roof, another proof". He is fond of idiosyncratic expressions (bosses, slaves and epsilons, captured and liberated, Sam and Joe, preaching, poison, cured of incurable disease), but the light-hearted, easy-going exterior hides a thoroughly professional approach to mathematics.

He has written more than 1200 papers with an astonishing number of co-authors – about 300. This phenomenon has given rise to the notion of an "Erdös number": even Einstein has an Erdös number, namely 2, and the largest Erdös number is believed to be 7.

This volume consists of research papers from diverse areas of mathematics, with a spread reflecting the wide range of Paul Erdös's interests. We are delighted to present it to him for his seventy-fifth birthday, on behalf of his many friends, collaborators and admirers, in appreciation of all that he has done for mathematics and the mathematical community during his long and fruitful career.'

Alan Baker

Poincaré's sudden illumination

'Just at this time, I left Caen, where I was living, to go on a geologic excursion under the auspices of the School of Mines. The incidents of the travel made me forget my mathematical work. Having reached Coutances, we entered an omnibus to go to some place or other. At the moment when I put my foot on the step, the idea came to me, without anything in my former thoughts seeming to have paved the way for it, that the transformations I had used to define the Fuschian functions were identical with those on non-Euclidian geometry. I did not verify the idea; I should not have had time, as, upon taking my seat in the omnibus, I went on with a conversation already commenced, but I felt a perfect certainty. On my return to Caen, for conscience's sake, I verified the results at my leisure.'

Saunderson, a blind mathematician

Nicholas Saunderson (1682–1739) was blinded by smallpox in his twelfth year. Nevertheless, amazing to relate, he was appointed in 1711 to Newton's chair at Cambridge, becoming the fourth Lucasian Professor of Mathematics.

The following account is from Diderot's *Letter on the Blind*, which was prompted by a question posed by William Mollineux and referred to by Locke in his *Essay Concerning Human Understanding*: would a blind man, on recovering his sight and being presented with a cube and globe before his eyes, be able to correctly name them?

'This Saunderson, Madame, is another blind person whom it would not be irrelevant to consider here. Wondrous things are told about him, and his progress in literature and in the mathematical sciences lends credence to all of them.

He was the author of a very perfect book of its kind, the *Elements of Algebra*, in which the only clue to his blindness is the occasional eccentricity of his demonstrations, which would perhaps not have been thought up by a sighted person. To him belongs the division of the cube into six equal pyramids having their vertices at the centre of the cube and the six faces as their bases; this is used for an elegant proof that a pyramid is one-third of a prism having the same base and height.

Saunderson taught mathematics at the University of Cambridge with astonishing success. He gave lessons in optics, and on the nature of light and colours; he explained the theory of vision; he considered the effects of lenses, the rainbow and many other matters relating to sight and the eye. These facts lose much of their strangeness, Madame, if you consider that there are three things which must be distinguished in any question that combines geometrical and physical considerations: the phenomena to be explained; the axioms of the geometry; and the calculation which follows from the axiom. Now, it is obvious that however acute the blind man may be, the phenomena of light and colour are completely unknown to him. He will understand the axioms, because he refers them to palpable objects, but he will not understand why geometry should prefer them to other axioms, for to do so he would have to compare the axioms with the phenomena directly, which for him is an impossibility. The blind man thus takes the axioms as they are given to him; he interprets a ray of light as a thin elastic thread, or as a succession of tiny bodies that strike the eyes with incredible force – and he calculates accordingly. The boundary between physics and mathematics has been crossed, and the problem becomes purely formal.'

Saunderson invented the 'pin-board'. His version is illustrated overleaf. It consisted of many sets of nine holes, each arranged in three rows of three, into which small pegs fitted. When he used this aid for arithmetical calculation, at which he became extraordinarily proficient, each hole stood for a digit. When used as a geometrical aid, he joined the pegs with thread to form the figures. A similar, albeit simpler, device is nowadays used by schoolchildren as an aid to geometry.

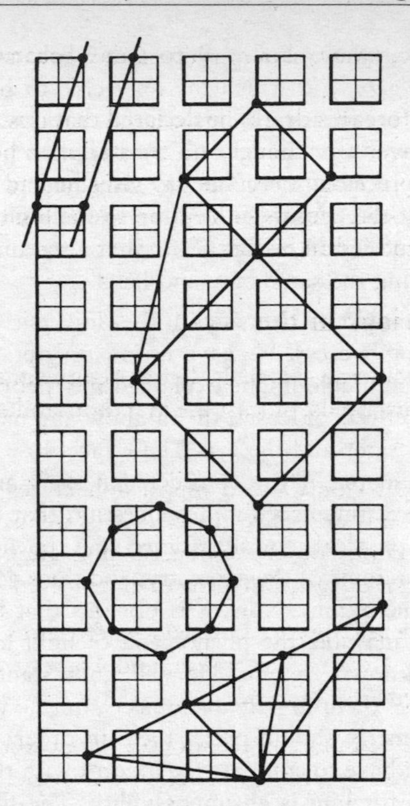

De Moivre approaches the limit

Of Abraham de Moivre (1667–1754), a Huguenot who fled to London in 1688 following the revocation of the Edict of Nantes, it was said that 'in the long list of men ennobled by genius, virtue and misfortune, who have found asylum in England, it would be difficult to name one who has conferred more honour on his country'.

He is most famous for De Moivre's formula, the beautiful and profound result that $(\cos x + i\sin x)^n = \cos nx + i\sin nx$ [where $i = \sqrt{-1}$]. Among his other achievements, he was one of the commissioners appointed by the Royal Society in 1712 to judge between the claims of Newton and Leibniz to have invented the infinitesimal calculus. He was also the unique example of a professional at chess and mathematics.

'In his old age, twenty hours' sleep a day became habitual with De Moivre.

"Shortly before [his death] he declared that it was necessary for him to sleep some ten minutes or a quarter of an hour longer each day than the preceding one. The day after he had reached a total of over twenty-three hours he slept up to the limit of twenty-four hours and then died in his sleep." '

Mathematicians in the sky

Gaussia (asteroid 1001) is named after Gauss, an especial honour because most asteroids, at least the first thousand or so, have been given Greek names.

Many Greek astronomers who appear in histories of mathematics as mathematicians appear in the sky. Aristarchus is the name of the brightest crater on the moon, named after Aristarchus of Samos who first suggested that the earth revolved around the sun.

Halley's comet last appeared in 1985, and returns every 76 years. It is named, of course, after Edmund Halley, who persuaded Newton to write his *Principia*.

There is a Triangulum constellation in the northern hemisphere, and a Triangulum Australe, or Southern Triangle, in the southern hemisphere.

There is a crater on the moon, named McMath, after not one but two American astronomers. More seriously, more than seventy moon craters are named after mathematicians, from Abel, Abul Wáfa, Archimedes and Babbage, to Dirichlet, Euclid, Euler, Fermat, Galois and Gauss, to Newton, Emmy Noether and Omar Khayyám, to Vieta, Von Neumann, Weyl and Wiener.

The death of John von Neumann

'In the summer of 1955 von Neumann slipped on a corridor in an office building and hurt his left shoulder. The injury led to the diagnosis of a bone cancer in August. It had already metastasized. His first response to the situation was optimistic, even stoic, and he worked harder than ever. To a remarkable extent he continued to meet the many demands made on him in his capacity as AEC commissioner and to pursue his other interests. There was a bitter irony in the situation in that von Neumann, who had waved away the cancer-producing effects of nuclear weapons tests, would himself contract this awful disease. He had increased his own chances of getting cancer by personally attending nuclear weapons tests and by staying at Los Alamos for long periods.

Before long his accustomed four or five hours of sleep did not suffice, and his enormous activity was forced into a slower pace. Yet death was unthinkable, inconceivable – so many plans and projects were still unfulfilled. The psychological stress was enormous, the more so as his special sense of invulnerability, which many of us share but to a lesser degree, was being challenged. To ease his spiritual troubles von Neumann sought not a Jewish Rabbi but a priest to instruct him in the Catholic faith. Morgenstern, among others, was shocked. "He was of course completely agnostic all his life, and then he suddenly turned Catholic – it doesn't agree with anything whatsoever in his attitude, outlook and thinking when he was healthy."

But religion did not prevent him from suffering; even his mind, the amulet on which he always had been able to rely, was becoming less dependable. Then came complete psychological breakdown; panic; screams of uncontrollable terror every night. His friend Edward Teller said, "I think that von Neumann suffered more when his mind would no longer function, than I have ever seen any human being suffer."

Von Neumann's sense of invulnerability, or simply the desire to live, was struggling with unalterable facts. He seemed to have a great fear of death until the last ... No achievements and no amount of influence could save him from extinction now, as they always had in the past. Johnny von Neumann, who knew how to live so fully, did not know how to die.

Physiologically his brain was fully intact, not touched by the cancer, but he had been suffering steadily from physical pain. And still the United States government depended on his thinking.

... von Neumann was assigned an aide, Air Force Lt. Col. Vincent Ford, and air force hospital orderlies with top-secret security clearance, lest in his distraction he should babble "classified information". He died February 8, 1957, at the age of fifty-three.'

Einstein's regrets

'The book [the poetry of Virgil] shows me clearly what I fled from when I sold myself body and soul to Science – the flight from the I and WE to the IT.'

Einstein

A free slice of pi

'There is a story about some American legislature having considered a bill to legislate, for religious reasons, the biblical value of $\pi = 3$. I have found no confirmation of this story; very probably it grew out of an episode that actually took place in the State Legislature of Indiana in 1897. The Indiana House of Representatives did consider and unanimously pass a bill that attempted to legislate the value of π (a wrong value); the author of the bill claimed to have squared the circle, and offered this contribution as a free gift for the sole use of the State of Indiana (the others would evidently have to pay royalties).

The author of the bill was a physician, Edwin J. Goodwin, MD, of Solitude, Posey County, Indiana, and it was introduced in the Indiana House on January 18, 1897, by Mr Taylor I. Record, Representative from Posey County. It was entitled "A bill introducing a new Mathematical truth", and it became House Bill No. 246; copies of the bill are preserved in the Archives Division of the Indiana State Library . . .

The bill was, perhaps symbolically, referred to the House Committee on Swamp Lands, which passed it on to the Committee of Education, and the latter reported it back to the House "with recommendation that said bill do pass". On February 5, 1897, the House passed the learned treatise unanimously (67 to 0).

Five days later the bill went to the Senate, where it was referred, for unknown reasons, to the Committee on Temperance. The Committee on Temperance, too, reported it back to the Senate with the recommendation that it pass the bill, and it passed the first reading without comment.

What would have happened to the bill if events had run their normal course is anybody's guess. But it so happened that Professor C. A. Waldo, a member of the mathematics department of

Purdue University, was visiting the State Capitol to make sure the Academy appropriation was cared for, and as he later reported, was greatly surprised to find the House in the midst of a debate on a piece of mathematical legislation; an ex-teacher from eastern Indiana was saying: "The case is perfectly simple. If we pass this bill which establishes a new and correct value of π, the author offers our state without cost the use of this discovery and its free publication in our school textbooks, while everyone else must pay him a royalty."

Professor Waldo, horrified that the bill had passed the House, then coached the senators, and on its second reading, February 12, 1897, the Senate voted to postpone the further consideration of this bill indefinitely; and it has not been on the agenda since.'

Petr Beckmann

Pure mathematics

'A headmistress in Thatcher, Arizona, decided to remove all controversial books from the library and when the purge was over nothing remained on the shelves but mathematics textbooks.'

The Professor

It was George Polya who admitted to studying mathematics at college because physics was too hard and philosophy was too easy. This is his view of the traditional mathematics professor.

'The traditional mathematics professor of the popular legend is absentminded. He usually appears in public with a lost umbrella in each hand. He prefers to face the blackboard and to turn his back on the class. He writes *a*, he says *b*, he means *c*; but it should be *d*. Some of his sayings are handed down from generation to generation.

"In order to solve this differential equation you look at it till a solution occurs to you."

"This principle is so perfectly general that no particular application of it is possible."

"Geometry is the art of correct reasoning on incorrect figures."

"My method to overcome a difficulty is to go round it."

"What is the difference between method and device? A method is a device which you use twice."

After all, you can learn something from this traditional mathematics professor. Let us hope that the mathematics teacher from whom you cannot learn anything will not become tradi tional.'

Reason versus Imagination

'I have often been surprised that Mathematics, the quintessence of Truth, should have found admirers so few and so languid. Frequent consideration and minute scrutiny have at length unravelled the cause: viz. that though Reason is feasted, Imagination is starved; whilst reason is luxuriating in its proper Paradise, Imagination is wearily travelling on a dreary desert. To assist Reason by the stimulus of Imagination is the design of the following production.'

> Samuel Taylor Coleridge, in a preface to one of his earliest poems, in which he rendered the first book of Euclid into verse.

Coleridge studied mathematics at Cambridge, and the very first two entries in his private notebooks were mathematical puzzles, but he clearly lacked Voltaire's appreciation of mathematics (see page 168).

Gulliver in Lagado

Jonathan Swift (1667–1745) was antagonistic to the Royal Society and to the 'new philosophy' which he satirized in *Gulliver's Travels*.

'The knowledge I had in mathematics gave me great assistance in acquiring their phraseology, which depended much upon that science and music; and in the latter I was not unskilled. Their ideas are perpetually conversant in lines and figures. If they would, for example, praise the beauty of a woman or any other animal, they describe it by rhombs, circles, parallelograms, ellipses, and other geometrical terms, or else by words of art drawn from music, needless here to repeat. I observed in the King's kitchen all sorts of mathematical and musical instruments, after the figures of which they cut up the joints that were served to his Majesty's table.

Their houses are very ill built, the walls bevil, without one right angle in any apartment; and this defect ariseth from the contempt they bear for practical geometry, which they despise as vulgar and mechanic, those instructions they give being too refined for the intellectuals of their workmen, which occasions perpetual mistakes. And although they are dextrous enough upon a piece of paper in the management of the rule, the pencil, and the divider, yet in the common actions and behaviour of life I have not seen a more clumsy, awkward, and unhandy people, nor so slow and perplexed in their conceptions upon all other subjects, except those of mathematics and music. They are very bad reasoners, and vehemently given to opposition, unless when they happen to be of the right opinion, which is seldom their case. Imagination, fancy, and invention, they are wholly strangers to, nor have any words in their language by which those ideas can be expressed; the whole compass of their thoughts and mind being shut up within the two forementioned sciences.

I was at the mathematical school, where the master taught his pupils after a method scarce imaginable to us in Europe. The proposition and demonstration were fairly written on a thin wafer, with ink composed of a cephalic tincture. This the student was to swallow upon a fasting stomach, and for three days following eat nothing but bread and water. As the wafer digested, the tincture mounted to his brain, bearing the proposition along with it. But the success hath not hitherto been answerable, partly by some error in the *quantum* or composition, and partly by the perverseness of lads, to whom this bolus is so nauseous that they generally steal aside, and discharge it upwards before it can operate; neither have they been yet persuaded to use so long an abstinence as the prescription requires.'

A notable failure

'Einstein once invited the renowned pianist A. Schnabel for a musical week-end. They were running through a rather intricate Mozart sonata, and Einstein was having trouble playing. Finally, after several explanations, Schnabel became irritated. He banged his hands down on the keyboard and groaned: "No, no, Albert. For heaven's sake, can't you count? One, two, three, four . . ." '

God

Plato: God ever geometrizes!

Jacobi: God ever arithmetizes!

Kronecker: God created the natural numbers, all else is the work of man!

When Henry Briggs (1561–1630) died, his epitaph claimed that 'his soul still astronomizes and his body geometrizes'.

An Arabian first

Early mathematicians tended to be interested in finding *a solution* to a problem, rather than finding *all possible* solutions. Abu Kamil was one of the first, if not the first, to take seriously the problem of counting the number of solutions, in his *Book of Arithmetical Rarities*, written about AD 900.

' "In the Name of God, the Compassionate, the Merciful: these are the words of Shodja ben Aslam, known as Abu Kamil. I am acquainted with a type of problem which proves to be engrossing, novel and attractive alike to high and low, to the learned and to the ignorant. But when others discuss solutions with one another, they exchange inaccuracies and conjectures, as they see no evident principle or system.

"Many – both high and low – have kept bringing me such questions; and I have given them the unique answer, when there was one: but often there were two, three, four or more answers, and often none. Once, indeed, a problem was brought to me, and I solved it, obtaining very many solutions; I went into it fully, and found that there were 2,678 valid answers. I marvelled at this, only to discover – when I spoke of it – that I was reckoned a simpleton or an incompetent, and strangers looked on me with suspicion.

"So I decided to write a book, to make the matter better understood. This is its beginning. I shall show the working for problems which have several solutions, for problems which have only one, and for problems which have none at all. And after this I shall deal with the problem for which – as I said – there are 2,678 solutions.

"This will sweep away the calumnies and conjectures. I shall justify my assertions, and truth will become evident. The book would become too long if I added everything suggested to me by the large number of solutions of this and similar problems." '

The problem he considered led to the pair of indeterminate equations,

$$x + y + z + u + v = 100$$
$$2x + \tfrac{1}{2}y + \tfrac{1}{3}z + \tfrac{1}{4}u + v = 100$$

for which he sought solutions in positive integers. One such solution is $x = 31$, $y = 18$, $z = 6$, $u = 24$, $v = 21$.

The Importance of Form

'It may surprise some people to learn that a psychological test designed to detect in artists preference for good form and design is: "the most powerful single test yet discovered as a predictor of creative potential in any field of endeavour . . . One would not be surprised to see such a test correlating with creativity in the arts and perhaps in literature but we need to note that the test is an equally good predictor or creative potential in the physical sciences and engineering." '

Anthony Storr

And, we might add, in mathematics. Without appreciation of form, there can be no feeling for analogy. Without perception of analogy, there can be no conception of structure. Without structure, mathematics would only be a collection of techniques, mundanely useful, but ultimately 'without form and void'.

Littlewood took this one stage further:

'I constantly meet people who are doubtful, generally without due reason, about their potential capacity [to appreciate mathematics]. The first test is whether you got anything out of geometry.'

Maxwell's analogies

Faraday was notable for his brilliant discoveries despite his lack of mathematics. Maxwell, who developed Faraday's theories, was a brilliant mathematician who, 'was unequalled in his skill at finding and using analogies between physical phenomena, not merely in illustrative roles, but as genuine research tools. His analogies were sometimes substantive, as with the "billiard ball" model for kinetic theory, and sometimes they were fragile and unreal, as in his mechanical development of electromagnetic theory . . . these early papers . . . show Maxwell developing a strong geometric sense, a valuable pre-requisite to physical imagery and modelling.'

The lack of an emphasis on such thinking is damaging:

'Unfortunately, there has recently developed a resurgence of the rather sterile formalistic attitude in theoretical studies, especially in applied mathematics, in which mechanisms are ignored and the deductive aspect given supreme status, akin to the Bourbakian style dominant in pure mathematics . . . it is the neglect of physical intuition, model making, geometrical interpretation, etc. as being outside an axiomatic based logical structure that is the real objection to formalism . . . (The atrophy of geometry from mathematics syllabuses in recent years is in phase with the growth of a puritanical formalism that is corroding applied mathematics.)'

L. C. Woods

A Valentine

My cardioid I sent to thee
With love and osculations:
O conjugate, let's have no more
Harmonic separations.
Love's Waves abound this day, and I
Wait thy reciprocations.

J.G.B.

A moral tale

There have been many claims for horses that could count, but Sugar left them all far behind. He could count to one hundred and also add, subtract, multiply and divide. Unfortunately, his master decided to try him on some elementary analytic geometry, which so frightened him that he ran off and was never seen again.

Moral: never put Descartes before the horse.

The landscape of mathematics

'It is difficult to give an idea of the vast extent of modern mathematics. This word "extent" is not the right one: I mean extent crowded with beautiful details – not an extent of mere uniformity such as an objectless plain, but of a tract of beautiful country seen at first in the distance, but which will bear to be rambled through and studied in every detail of hillside and valley, stream, rock, wood, and flower . . .'

Arthur Cayley (1821–1895)

Oliver Wendell Holmes on mathematics

'All economical and practical wisdom is an extension or variation of the following arithmetical formula: $2 + 2 = 4$. Every philosophical proposition has the more general character of the expression $a + b = c$. We are mere operatives, empirics and egotists, until we learn to think in letters instead of figures.

What a satire, by the way, is that (Babbage's calculating machine) on the mere mathematician! A Frankenstein monster, a thing without brains and without heart, too stupid to make blunders; that turns out formulae like a cornsheller and never grows any wiser or better, though it grind a thousand bushels of them!

I have immense respect for a man of talent *plus* "the mathematics". But the calculating power alone should seem to be the least human of qualities and to have the smallest amount of reason in it; since a machine can be made to do the work of three or four calculators and better than any one of them. Sometimes I have been troubled that I had not a deeper intuitive apprehension of the relations of numbers. But the triumph of the ciphering hand-organ has consoled me. I always fancy I can hear the wheels clicking in a calculator's brain. The power of dealing with numbers is a kind of "detached lever" arrangement, which may be put into a mighty poor watch. I suppose it [having "the mathematics"] is about as common as the power of moving the ears voluntarily, which is a moderately rare endowment.'

The roots of music

In 1988, Nicolas Slonimsky, aged 94, the world-renowned lexicographer of music, and conductor – he invented a method of beating a different rhythm with each arm – created a new composition by identifying each note in Beethoven's Fifth Symphony with a number, and then playing the square root of each note.

A source of disappointment

'. . . there is an amazingly high consensus in mathematics as to what is "correct" or "accepted". But alongside this, equally important, is the issue of what is "interesting" or "important" or "deep" or "elegant". These esthetic or artistic criteria vary widely, from person to person, speciality to speciality, decade to decade. They are perhaps no more objective than esthetic judgements in art or music.'

Reuben Hersh

Moser and Mordell

'[This toast] was sent to me by Professor L. Moser. Of him, it was said that he was writing a book and taking so long about it that his publishers became very much worried and went to see him. He said he was very sorry about the delay, but he was afraid that the book might have to be a posthumous one. Well, he was told, please hurry up with it.

Moser's toast was as follows:

Here's a toast to L. J. Mordell,
young in spirit, most active as well,
He'll never grow weary,
of his love, number theory,
The results he obtains are just swell.'

Julia Robinson and Hilbert's Tenth Problem

Julia Robinson and Martin Davis were responsible for most of the progress, made over a period of more than forty years, in attempts to solve Hilbert's Tenth Problem. This problem demanded a procedure that would determine in a finite number of steps whether a Diophantine equation (a polynomial equation with integer coefficients such as $3x^2 - 5y^2 = 2$) had a solution in rational numbers. The solution turned out to be that such a procedure does not exist. One spin-off from this negative but powerful conclusion was the 'formula' for the primes on page 146.

'Throughout the 1960s, while publishing a few papers on other things, I kept working on the Tenth Problem, but I was getting rather discouraged. For a while I ceased to believe in the Robinson hypothesis, although Raphael [her husband] insisted that it was true but just too difficult to prove. I even worked in the opposite direction, trying to show that there was a positive solution to Hilbert's problem, but I never published any of that work. It was the custom in our family to have a get-together for each family member's birthday. When it came time for me to blow out the candles on my cake, I always wished, year after year, that the Tenth Problem would be solved – not that I would solve it, but just that it would be solved. I felt that I couldn't bear to die without knowing the answer.

Finally – on February 15, 1970 – Martin telephoned me from New York to say that John Cocke had just returned from Moscow with the report that a 22-year-old mathematician in Leningrad had proved that the relation $n = F_{2m}$, where F_{2m} is a Fibonacci number, is diophantine. This was all that we needed. It followed that the solution to Hilbert's tenth problem is negative – a general method for determining whether a given diophantine equation has a solution in integers does not exist.

Just one week after I had first heard the news from Martin, I was able to write to Matijasevič:

"... now I know it is true, it is beautiful, it is wonderful.

"If you really are 22 [he was], I am especially pleased to think that when I first made the conjecture you were a baby and I just had to wait for you to grow up!"

That year when I went to blow out the candles on my cake, I stopped in mid-breath, suddenly realizing that the wish I had made for so many years had actually come true.

I have been told that some people think that I was blind not to see the solution myself when I was so close to it. On the other hand, no one else saw it either. There are lots of things, just lying on the beach as it were, that we don't see until someone else picks one of them up. Then we all see that one.

In 1971 Raphael and I visited Leningrad and became acquainted with Matijasevič and with his wife, Nina, a physicist. At that time, in connection with the solution of Hilbert's problem and the role played in it by the Robinson hypothesis, Linnik told me that I was the second most famous Robinson in the Soviet Union, the first being Robinson Crusoe.'

From numbers to God

'The true foundation of theology is to ascertain the character of God. It is by the aid of Statistics that law in the social sphere can be ascertained and codified, and certain aspects of the character of God thereby revealed. The study of statistics is thus a religious service.'

Florence Nightingale

Ramanujan's – what's the word?

Ramanujan was extremely religious, even mystical. One day, while explaining a mathematical point to an Indian friend, he suddenly turned to him and exclaimed, 'Sir, an equation has no meaning for me unless it expresses a thought of GOD.' It is, perhaps, natural, albeit inexplicable, that the creation of his mathematics was linked to his religious experiences.

'I had known Ramanujan when he was a boy in Kumbakonam. Before 1911, I was Assistant Account General in Madras for a few years. I was living in a bungalow in Purushawakkam. Ramanujan was then unemployed. He came to me one evening and wanted to sleep in my house. But all the rooms in the main building were occupied. There was only one outhouse. A monk was occupying it. I asked Ramanujan whether he would mind sleeping in the outhouse along with the monk. He welcomed it. Next morning he came and told me, "He is such a powerful soul, I am glad I had the opportunity to share the room with him. His presence stimulated me a good deal. While asleep I had an unusual experience. There was a red screen formed by flowing blood as it were. I was observing it. Suddenly a hand began to write on the screen. I became all attention. That hand wrote a number of results in elliptic integrals. They stuck to my mind. As soon as I woke up, I committed them to writing." '

T. K. Rajagopalan

'I am giving here an extract from my book *Hidden Treasures of Yoga*, which relate to Ramanujan. 'Ramanujan and his family were ardent devotees of God Narasimha (the lion-faced incarnation (*avasara*) of God), the sign of whose grace consisted in drops of blood seen during dreams. Ramanujan stated that after seeing such drops, scrolls containing the most complicated mathematics used to unfold before him and that after waking, he could set down on paper only a fraction of what was shown to him.'

R. Srinivasan

God's Marvelous Mathematical Book

'If I see a really nice proof, I say it comes straight from the Book . . . God has a transfinite Book, which contains all theorems and their best proofs, and if He is well intentioned toward those [mathematicians]. He shows them the Book for a moment. And you wouldn't even have to believe in God, but you must believe that the Book exists.'

Paul Erdös

The most prolific mathematician of all time

'Euler was so prolific that an entire volume is required to contain the list of his publications. Approximately one third of the entire corpus of research on mathematics and mathematical physics and engineering mechanics published in the last three-quarters of the eighteenth century is by Euler. From 1729 onward he filled about half of the pages of the publications of the Petersburg Academy, not only until his death in 1783 but on and on over fifty years afterward. (Surely a record for slow publication was won by the memoir presented by him to that academy in 1777 and published by it in 1830.) From 1746 to 1771 Euler filled approximately half of the scientific pages of the proceedings of the Berlin Academy also. He wrote for other periodicals as well, but in addition he gave some of his papers to booksellers for issue in volumes consisting wholly of his work. By 1910 the number of his publications had reached 866, and five volumes of his manuscript remains, a mere beginning, have been printed in the last ten years. There is almost no duplication of material from one paper to another in any one decade, and even most of his expository books, some twenty-five volumes ranging from algebra and analysis and geometry through mechanics and optics to philosophy and music, include results he had not published elsewhere. The modern edition of Euler's collected works was begun in 1911 and is not yet quite complete; although mainly limited to republication of works which had been published at least once before 1910, it will require about seventy-five large quarto volumes, each containing 300 to 600 pages. Euler left behind him also 3,000 pages of clearly and consecutively written mathematical notebooks and early draughts of several books. A whole volume is filled by the catalogue of the manuscripts preserved in Russia.'

Modern Mathematics for T. C. Mits, The Celebrated Man in the Street

VII ABSTRACTION

this power to ABSTRACT is
one of the outstanding characteristics
of human beings as
compared with other animals.
And this power is used not only
by mathematicians,
but also by
artists, musicians, poets,
and all other 'human' beings.
Perhaps some day
we shall measure
a person's 'human-ness' by
his power to abstract
rather than by the I.Q.
For a person who can
be loyal to such
abstract concepts as
truth, justice, freedom, reason,
rather than to
an individual or a place,
has the loyalty of a human being
rather than that of a dog.
Please do not think that
we are using the word 'dog'
in a disparaging sense,
for they are very dear animals.
(Remember that you must not be
a Conclusion-Jumper!)
But still they are animals and
not human beings.

But what are
'Truth',
'Justice',
'Freedom',

'Reason',
etc.?
Do these words really mean anything?
And how can we be loyal to them
if their meaning is not clear?
Are they not just 'fakes',
invented so that
some people can make slaves of others
by fooling them with such
meaningless abstractions?
Now you will see,
when you have finished this
little book,
that these concepts
'Truth', 'Freedom', 'Reason', etc.,
will become much clearer when
we examine into what is meant by
'Mathematical Truth',
what kind of 'Freedom' we have
in Mathematics,
what is considered good 'Reason'
in Mathematics,
and so on.

You will see that
as mathematicians have been
gradually forced to consider
the fundamentals of Mathematics,
they have been obliged
to consider the very nature
of human thinking –
both its powers and
its limitations.
For instance,
what is the nature of a 'proof'
by human beings
for human beings?

And, of course,
this has a definite bearing on:

'What are we humans anyway?
What is the best that
we can expect of ourselves?'

The Moral: Be a man – not a mouse.

Lilian R. Lieber

Dirac on pretty mathematics

'A good deal of my research work in physics has consisted in not setting out to solve some particular problem, but simply examining mathematical quantities of a kind that physicists use and trying to fit them together in an interesting way regardless of any application that the work may have. It is simply a search for pretty mathematics. It may turn out later that the work does have an application. Then one has had good luck.

I can give a good example of this procedure. At one time, in 1927, I was playing around with three 2×2 matrices whose squares are equal to unity and which anticommute with one another. Calling them σ_1, σ_2, σ_3, I noticed that if one multiplied them into the three components of a momentum so as to form $\sigma_1 p_1 + \sigma_2 p_2 + \sigma_3 p_3$, one obtained a quantity whose square was just $p_1^2 + p_2^2 + p_3^2$. This was an exciting result, but what use could one make of it?

One could use $\sigma_1 p_1 + \sigma_2 p_2 + \sigma_3 p_3$ as the Hamiltonian in a Schrödinger wave equation, giving the wave function two components so that the σ matrices can be applied to it. One then had a relativistic wave equation. But it applied only to a particle of zero rest mass. To get a theory for a particle with nonzero rest mass one would need four σ matrices anticommuting with one another, and such matrices did not exist. So my work was of no use for the electron, which was what I was mainly interested in. I therefore had to abandon it.

It was not until some weeks later that I realized there is no need to restrict oneself to 2×2 matrices. One could go on to 4×4 matrices, and the problem is then easily soluble. In retrospect, it seems strange that one can be so much held up over such an elementary point.

The resulting wave equation for the electron turned out to be very successful. It led to correct values for the spin and the magnetic moment. This was quite unexpected. The work all followed from a study of pretty mathematics, without any thought being given to these physical properties of the electron.'

A Sampler of Statistics

'Statistics is like a bikini. What they reveal is suggestive, what they conceal is vital.'

Aaron Levenstein

'We must believe in luck, otherwise how can we explain the success of those we don't like?'

Jean Cocteau

'Life is the art of drawing sufficient conclusions from insufficient premises.'

Samuel Butler

'To guess is cheap, to guess wrongly is expensive.'

Old Chinese proverb

' "Give us a copper, Guv", said the beggar to the Treasury statistician, when he waylaid him in Parliament Square. "I haven't eaten for three days." "Ah," said the statistician, "and how does that compare with the same period last year?" '

Russell Lewis

'There are three kinds of lies: lies, damned lies and statistics.'

Mark Twain

'I can prove anything by statistics except the truth.'

George Canning

'He uses statistics as a drunk uses a street lamp, for support rather than illumination.'

Andrew Lang

'A statistician is a person who draws a mathematically precise line from an unwarranted assumption to a foregone conclusion.'

Anon.

Carlyle as a mathematician

Thomas Carlyle, the Sage of Chelsea and historian of the French Revolution, studied mathematics at Edinburgh University. His teacher, Leslie, had a high opinion of his mathematical ability and helped him to get his first appointment as a mathematics teacher.

In 1822 he completed his translation of Legendre's *Elements of Geometry and Trigonometry*, for which 'I got only £50 for my entire trouble . . . and had already ceased to be the least proud of my mathematical prowess'.

This edition was 'revised and adapted to the course of mathematical instruction in the United States' by Charles Davies, and went through nearly thirty editions between 1834 and 1890, becoming the leading text book on its subject.

Ironically, his literary work called down the scorn of scientists. Darwin remarked, 'I never met a man with a mind so ill-adapted for scientific research,' and Spencer thought him 'sloppy'. Yet de Morgan praised his mathematical work, especially an essay on proportion which was, 'as good a substitute for the fifth book of Euclid as could be given in the space'.

Two years later Carlyle met Legendre, and attended a meeting of the Institute where he saw Laplace and Poisson, but his interest in mathematics was waning, and he never returned to it.

Counting-out rhymes

Eeny, meeny, miney, mo,
Sit the baby on the po,
When he's done
Wipe his bum,
Tell his mummy what he's done.

Inky, pinky, ponky,
My dad bought a donkey,
The donkey died,
Daddy cried,
Inky, pinky, ponky.

Children have used counting-out rhymes to pick out a special child throughout history. One child goes rapidly round the group, pointing to a child as each word is said. The last child becomes the chaser, for example, or is eliminated and the counting starts again, until eventually only a single child is left.

The rhymes frequently resemble limericks, and start with nonsense words, both of which children love, such as 'Eeny, meeny, miney, mo', or 'Hickety pickety i sillickety', or 'Eenty-peenty, halligo lum'. But are these mere nonsense, or do they have a deeper origin?

The old verse,

> Hickory, dickory, dock,
> The mouse ran up the clock,
> The clock struck one,
> The mouse ran down,
> Hickory, dickory, dock.

is recorded as a nursery rhyme, rather than a counting-out rhyme, (and is clearly of limerick form). The first line, however, is not nonsense, but a corruption of the 'shepherd's score, so called, the numerals reputedly employed in past times by shepherds counting their sheep, by fishermen assessing their catch, and old women minding their stitches'. They are found throughout the north of England. This is how Melvyn Bragg used them in his novel, *A Place in England*:

'Joseph rhymed off the count from one to twenty, in his own, the West Cumbrian dialect, singing it almost, as the words demanded:

"Yan, tyan, tethera, methera, pimp, sethera, lethera, hovera, dovera, dick. Yan-a-dick, tyan-a-dick, tethera-dick, methera-dick, bumfit."

"Bumfit!" Mr Lenty interrupted ecstatically. "Oh, thou Bumfit! My Bumfit! Now why can't we still say Bumfit. Fifteen doesn't hold a candle to it. Bumfit! – go on, Joseph."

"Yan-a-bumfit, tyan-a-bumfit, tethera-bumfit, methera-bumfit, giggot."

"Giggot!" said Mr Lenty. "Twenty. And-the-days-of-thy-years-are-tethera-giggots-and-dick. Now isn't that better than three score and ten? It sounds like a lifetime, doesn't it? I could hear you repeat that all evening." '

Melvyn Bragg

The limits of imagination

'I have dwelt the longer on this subject, because I think it may show us the proper limits, as well as the defectiveness of our imagination; how it is confined to a very small quantity of space, and immediately stopped in its operation, when it endeavours to take in any thing that is very great or very little. Let a man try to conceive the different bulk of an animal, which is twenty, from another which is a hundred times less than a mite, or to compare in his thoughts a length of a thousand diameters of the earth, with that of a million; and he will quickly find that he has no different measures in his mind, adjusted to such extraordinary degrees of grandeur or minuteness. The understanding, indeed, opens an infinite space on every side of us; but the imagination, after a few faint efforts, is immediately at a stand, and finds herself swallowed up in the immensity of the void that surrounds it: our reason can pursue a particle of matter through an infinite variety of divisions; but the fancy soon loses sight of it, and feels in itself a kind of chasm, that wants to be filled with matter of a more sensible bulk. We can neither widen nor contract the faculty to the dimensions of either extreme. The object is too big for our capacity, when we would comprehend the circumference of a world; and dwindles into nothing when we endeavour after the idea of an atom.'

Joseph Addison (1672–1719)

The pattern of mathematics

Among the words in the English language which have no rhymes are common, every, hundred, method, sausage, and sunrise and sunset, but not 'hypotenuse', thanks to W. S. Gilbert, who put some of the best rhymes, as well as the best tongue-twisters in the English language, into his patter songs. Here are the first three verses of the famous song sung by Major-General Stanley, from *The Pirates of Penzance*.

GEN. I am the very model of a modern Major-General,
I've information vegetable, animal, and mineral,
I know the kings of England, and I quote the fights
historical,

From Marathon to Waterloo, in order categorical;
I'm very well acquainted too with matters mathematical,
I understand equations, both the simple and quadratical,
About binomial theorem I'm teeming with a lot o' news –
With many cheerful facts about the square of the hypotenuse.

ALL. With many cheerful facts, etc.

GEN. I'm very good at integral and differential calculus,
I know the scientific names of beings animalculous;
In short, in matters vegetable, animal, and mineral,
I am the very model of a modern Major-General.

ALL. In short, in matters vegetable, animal, and mineral,
He is the very model of a modern Major-General.

GEN. I know our mythic history, King Arthur's and Sir Caradoc's,
I answer hard acrostics, I've a pretty taste for paradox,
I quote in elegiacs all the crimes of Heliogabalus,
In conics I can floor peculiarities parabolous.
I can tell undoubted Raphaels from Gerard Dows and Zoffanies,
I know the croaking chorus from the *Frogs* of Aristophanes,
Then I can hum a fugue of which I've heard the music's din afore,
And whistle all the airs from that infernal nonsense *Pinafore*.

Gilbert made a slight error, however, when he wrote in 1884, for the operetta *Princess Ida*, the lines,

As for fashion, they forswear it
So they say – so they say.
And the circle they will square it
Some fine day – some fine day.

Lindemann had proved that π was transcendental, from which it follows that the circle cannot be squared, in 1882.

Asymptotes

'Jacques Peletier was telling me at my house that he had found two lines approaching each other, which, however, he established could never succeed in meeting except at infinity.'

Michel de Montaigne (1533–1592)

Chilling mathematics

There is a common tendency among those who do not enjoy mathematics to describe it in terms such as 'cold', 'austere' and 'impersonal'. Thus Hogben wrote: 'the aesthetic appeal of mathematics may be very real for a chosen few ... [for those few] mathematics exercises a coldly impersonal attraction.'

Saunders Maclane refers to 'the view that mathematics is in part a search for austere forms of beauty'. Did Gian-Carlo Rota have the same feeling in mind when, while reviewing a book in *Advances in Mathematics*, he wrote: 'The chilling elegance of this presentation will give you goose pimples, it is hard to conceive of anything more beautiful or as deep'?

Serge Lang presumably meant the same when he answered questions from a public audience about mathematics, and in answer to the question, 'Why do you do this kind of work?' replied, 'Because it gives me chills in the spine.'

Chandrasekhar wrote, referring to Kerridge's solutions of Einstein's equations which describe a rotating black hole: 'This "shuddering before the beautiful", this incredible fact that a discovery motivated by a search for the beautiful in mathematics should find its exact replica in Nature, persuades me to say that beauty is that to which the human mind responds at its deepest and most profound. Chandrasekhar, in his book *Truth and Beauty*, also quoted Plato on the soul shuddering, suggesting that such feelings, whatever their psychological significance, are old.

I have left the best-known quote on this theme to the last. Bertrand Russell wrote that, 'Mathematics, rightly viewed, possesses not only truth, but supreme beauty – a beauty cold and austere, like that of sculpture, without appeal to any part of our weaker nature, without the gorgeous trappings of painting or music, yet sublimely pure, and capable of a stern perfection such as only the greatest art can show.'

Why mathematicians exist

'It must be recognised that the purely deductive method is wholly inadequate an instrument of research ... in research the psychological factor is of paramount importance ... The possession of a body of indefinables, axioms, or postulates, and symbols denoting logical relation, would, taken of itself, be wholly insufficient for the development of mathematical theory. With these alone the mathematician would be unable to move a step. In face of an unlimited number of possible combinations, a principle of selection of such as are of interest, a purposive element, and a perceptive faculty are essential for the development of anything new.'

Prof. E. W. Hobson

A comedy of errors

The development of the theory of groups shows numerous errors committed by eminent mathematicians.

'Galois stated in 1830 that the modular group can be represented as a substitution group of degree p, only when p = 5. Later he corrected this statement, to include p = 7 or 11, but no larger value.

Cauchy began the construction of fairly long lists of substitution groups which are supposed to contain all the possible substitution groups on six or a small number of letters. Such lists were based by Cauchy upon an insufficient knowledge of the theory of substitutions and hence they were incomplete.

Cayley's writings on group theory are scarcely more accurate than those of Cauchy ... Cayley states that there are three groups of order 6, giving as examples, the non-cyclic group and the cyclic group of order 6. It seems very strange that Cayley should have made a mistake regarding such a simple matter more than twenty years after he began to study group theory.

While the number of errors in [Jordan's] works is considerable it is perhaps not larger than one could reasonably expect from the difficulty and newness of the problems. In particular, as early as 1878 he made the first attempt to determine all the finite collineation groups in three variables, but he failed to find the two important groups of orders 168 and 360 . . .'

G. A. Miller

Gauss's precocity

'Carl was one of those remarkable infant prodigies who appear from time to time. They tell of him the incredible story that at the age of three he detected an arithmetical error in his father's bookkeeping. And there is the often-told story that when ten years old . . . his teacher, to keep the class occupied, set the pupils to adding the numbers from 1 to 100. Almost immediately Carl placed his slate, writing side down, on the annoyed teacher's desk. When all the slates were finally turned in, the amazed teacher found that Carl alone had the correct answer, 5050, but with no accompanying calculation. Carl had mentally summed the arithmetic progression $1 + 2 + 3 + \ldots + 98 + 99 + 100$ by noting that $100 + 1 = 101$, $99 + 2 = 101$, $98 + 3 = 101$, and so on for 50 such pairs, whence the answer is 50×101, or 5050. Later in life Gauss used to claim jocularly that he could figure before he could talk.'

Professional disagreements

Mathematicians have long recognized their differences in style from their colleagues. Jakob Steiner, a brilliant synthetic geometer, hated analytic methods in geometry. He felt that 'geometry could best be learned by concentrated thought, and he objected even to such "props" as the models and diagrams which synthetic geometers employed'.

This was unfortunate for Plucker: Steiner threatened to cease contributing to Crelle's journal, if it continued to accept papers from him. Fourier got into hot water with Jacobi in a similar manner:

'Author of the *Théorie analytique de la Chaleur*, Fourier had contributed most brilliantly both to the birth of mathematical physics and also to an essential advance in analysis: the introduction of trigonometric series. Because of his own tastes and the general trend of his work, he did not fully appreciate the importance of some mathematical studies which he considered to be purely theoretical. Such a point of view was legitimate, since each scientist has a marked predilection for a particular domain of research, but in reporting to the *Académie des Sciences* some of Abel and Jacobi's basic work on elliptic integrals, Fourier made the great mistake of trying to impose his personal taste on others, and of stating his regret that scientists of such worth should choose to spend their time on purely theoretical research, rather than on the solution of problems of mathematical physics. In a letter to Legendre, Jacobi replied to this remark with some indignation:

"It is true that M. Fourier believes that the chief aim of mathematics is its public usefulness and its explanation of natural phenomena, but a philosopher like him ought to have known that the sole aim of science is to do honour to the human spirit, and that in this respect a question about numbers is as important as a question about the system of the Universe."

The sudden death of Fourier put a stop to this polemic, but the fruitful influence of some of Fourier's results on pure mathematics, and equally the relevance of Jacobi's work to applied mathematics, show clearly that an originator can but rarely evaluate the subsequent repercussions of his own discoveries.'

The Mathematical theory of big game hunting

'In the 1930s a small group of mathematical wags banded together under the pseudonym E. S. Pondiczery, purportedly of the Royal Institute of Poldavia. The name and institution were chosen to fit the initials E.S.P., R.I.P. (standing for "extrasensory perception, rest in peace"), for the group had planned to write an article on extrasensory perception. The article never appeared in print.

Pondiczery's main interest lay in mathematical curiosa, and his best-known contribution to this field appeared in *The American Mathematical Monthly* (August–September 1938, pp. 446–7) under the title, "A contribution to the mathematical theory of big game hunting". Because of the facetious nature of this contribution, Pondiczery sought permission of editor-in-chief Elton

James Moulton of the *Monthly* to use the pseudonym H. Pétard. The permission was granted, giving rise to the only known instance in mathematical literature of a paper published under a pseudonym of a pseudonym.'

The basic puzzle is simply stated: 'There are lions in the Sahara desert. Devise methods for capturing them.' Since the problem was first posed, dozens of methods have been proposed. Here is a small selection.

The Hilbert Method. Place a locked cage in the desert. Set up the following axiomatic system.

(i) The set of lions is non-empty.

(ii) If there is a lion in the desert, then there is a lion in the cage.

Theorem 1: There is a lion in the cage.

The Method of Inversive Geometry. Place a locked, spherical cage in the desert, empty of lions, and enter it. Invert with respect to the cage. This maps the lion to the interior of the cage, and you outside it.

The Projective Geometry Method. The desert is a plane. Project this to a line, then project the line to a point inside the cage. The lion goes to the same point.

The Bolzano–Weierstrass Method. Bisect the desert by a line running N–S. The lion is in one half. Bisect this half by a line running E–W. The lion is in one half. Continue the process indefinitely, at each stage building a fence. The lion is enclosed by a fence of arbitrarily small length.

The Method of Parallels. Select a point in the desert and introduce a tame lion not passing through that point. There are three cases:

(a) The geometry is Euclidean. There is then a unique parallel lion passing through the selected point. Grab it as it passes.

(b) The geometry is hyperbolic. The same method will now catch infinitely many lions.

(c) The geometry is elliptic. There are no parallel lions, so every lion meets every other lion. Follow a tame lion and catch all the lions it meets: in this way every lion in the desert will be captured.

The Thom–Zeeman Method. A lion loose in the desert is an obvious catastrophe. It has three dimensions of control (2 for position, 1 for time) and one dimension of behaviour (being parametrized by a lion). Hence by Thom's Classification Theorem it is a swallowtail. A lion that has swallowed its tail is in no state to avoid capture.

The Eratosthenian Method. Enumerate all objects in the desert; examine them one by one; discard all those that are not lions. A refinement will capture only prime lions.

The Peano Method. There exists a space-filling curve passing through every point of the desert. It has been remarked that such a curve may be traversed in as short a time as we please. Armed with a spear, traverse the curve faster than the lion can move his own length.

Backward Induction. We prove by backward induction the statement L(n): 'It is possible to capture n lions.' This is true for sufficiently large n since the lions will be packed like sardines and have no room to escape. But trivially L(n + 1) implies L(n) since, having captured n + 1 lions, we can release one. Hence L(1) is true.

The Bourbaki Method. The capture of a lion in a desert is a special case of a far more general problem. Formulate this problem and find necessary and sufficient conditions for its solution. The capture of a lion is now a trivial corollary of the general theory, which *on no account should be written down explicitly.*

Surgery. The lion is an orientable 3-manifold with boundary and so may be rendered contractible by surgery. Contract him to Barnum and Bailey.

The Postnikov Method. The lion, being hairy, may be regarded as a fibre space. Construct a Postnikov decomposition. A decomposed lion must, of course, be long dead.

The Game-theory Method. The lion is big game, hence certainly a game. There exists an optimal strategy. Follow it.

Mozart's Musical Game

'In 1793, J. J. Hummel published "instructions in four languages on how to compose waltzes and counter-dances with two dice". In 1806, C. Wheatstone of London came out with an identical game under the title of "Mozarts Musical Game, fitted in an elegant box, showing by an easy system [how] to compose an unlimited number of waltzes, rondos, hornpipes and reels". Whether Mozart was the originator of this game remains an open question. Some of his notebooks suggest that he was at least seriously interested in such a game [. . .] Musicologists believe that Joseph Haydn and C. P. E. Bach also had something to do with developing this game.'

This is one version of the rules for 'Mozart's Musical Game':

INSTRUCTIONS

Showing How Anyone, even if He Is Not Musical and Understands Nothing of Composition, Can Compose Counter-Dances or Anglaises with 2 Dice

1. The capital letters A to H over the 8 columns of the number table stand for the 8 measures of each part of the dance. For example, A stands for the first measure, B for the second, C for the third, and so on. The numbers in the columns indicate the number of the measure in the music supplied.
2. The numbers 2 to 12 are the possible sums that can be rolled with two dice.
3. If, for example, a player rolls a 6 for the first measure of the first part of the dance, he then looks next to the number 6 in Column A and finds the number 105, which stands for measure 105 in the music table. The player now writes out this measure and so has the beginning of the dance. He then rolls for the second measure. If he rolls an 8, for example, he will look at the number next to 8 in column B. This number is 81. The player finds this measure in the music table and copies it next to the first one. He continues in the same manner for 8 rolls, which complete the first part of the dance. Then he puts down a repeat sign and goes on to the second part.

Nursery Verses

One's none
Two's some
Three's many
Four's a penny
Five's a little hundred

Magpies

One for sorrow, two for joy
Three for a girl, four for a boy,
Five for silver, six for gold
Seven for a secret ne'er to be told

Grace Hopper

'Retired Rear Admiral Grace M. Hopper, an internationally renowned computer programming pioneer, died on January 1, 1992. She was 85 years old. Her career in the US Navy spanned 43 years, from World War II to her retirement in 1986.

After graduating with a Ph.D. in mathematics from Yale in 1934 she taught for nearly a decade at Vassar, then entered the Naval Reserve in 1943. She was assigned to the Bureau of Ordnance Computation Project at Harvard University, where she worked with the computer pioneer Howard Aiken.

In 1949 she joined the Eckert–Mauchly Computer Corporation. The company, whose founders had developed the ENIAC, one of the world's first digital computers, was at that time building the first commercial computer, the UNIVAC I. Grace Hopper played an active role in the design and development of COBOL, the Common Business Oriented Language, for use on the UNIVAC. COBOL is still widely used today for business applications.

After retiring from the Naval Reserve in 1966, Grace Hopper was recalled a year later to work on standardizing the Navy's computer languages. She finally retired in 1986 at the grand old age of 80. She had been promoted to the rank of Rear Admiral by presidential appointment in 1983.

It was Grace Hopper who coined the term "bug" for anything that causes trouble in a computer. The first computer bug was actually a moth, discovered one night in 1945 in a Harvard computer. This is how Grace Hopper tells the story.

"Things were going badly. There was something wrong in one of the circuits. Finally, someone located the trouble spot, and, using ordinary tweezers, removed the problem, a two-inch moth. From then on, when anything went wrong with a computer, we said it had bugs in it." '

The smallest uninteresting number

It is a well known fact that there is no smallest uninteresting integer, because if that number existed, it would have precisely that interesting and remarkable property. The argument applies to other sets of objects, as Simon Hoggart illustrates:

'My colleagues and I used to have a running debate about who was the most boring man in Britain, who – we naturally assumed – must be an MP. We switched our search to the second most boring man in Britain, since the most boring would, in a perverse way, be quite interesting. He has since lost his seat.'

By the same line of argument, there cannot be an ultimate or 'most trivial' piece of mathematical trivia, so we merely claim that this is very trivial indeed: the names of the seven bridges over the Pregel river which feature in Euler's problem of the Bridges of Königsberg were the Krämer, Schmiede, Holz, Hohe, Honig, Köttel and Grüne.

Freeman Dyson versus Bourbaki

'Unfashionable mathematics [in contrast to Bourbakiste mathematics] is mainly concerned with things of accidental beauty, special functions, particular number fields, exceptional algebras, sporadic finite groups . . . They have a quality of strangeness, of unexpectedness.'

Freeman Dyson

From amoeba to mathematician

'A series of creatures might be constructed, arranged according to their diminishing interest in the immediate environment, which would begin with the amoeba and end with the mathematician. In pure mathematics the maximum of detachment appears to be reached; the mind moves in an infinitely complicated pattern, which is absolutely free from temporal considerations. Yet this very freedom – the essential condition of the mathematician's activity – perhaps gives him an unfair advantage. He can only be wrong – he can't cheat. But the metaphysician can. The problems with which he deals are of overwhelming importance to himself and the rest of humanity, and it is his business to treat them with an exactitude as unbiased as if they were some puzzle in the theory of numbers.'

Lytton Strachey

Babbage corrects Tennyson

'When Tennyson wrote "The Vision of Sin", Babbage read it. After doing so, it is said he wrote the following extraordinary letter to the poet:

"In your otherwise beautiful poem, there is a verse which reads:

'Every moment dies a man,
Every moment one is born.'

"It must be manifest that, were this true, the population of the world would be at a standstill: In truth the rate of birth is slightly in excess of that of death. I would suggest that in the next edition of your poem you have it read:

'Every moment dies a man,
Every moment 1 1/6 is born.'

"Strictly speaking this is not correct. The actual figure is a decimal so long that I cannot get it in the line, but I believe 1 1/6 will be sufficiently accurate for poetry. I am, etc." '

Sylvester and Huxley, and structure

The standard nineteenth-century view contrasted mathematics as a deductive science modelled on Euclid with the inductive natural sciences. This was the great biologist T. H. Huxley's view, but the mathematician Joseph Sylvester (1814–1897) disagreed:

'[Huxley] says "mathematical training is almost purely deductive. The mathematician starts with a few simple propositions, the proof of which is so obvious that they are called self-evident, and the rest of his work consists of subtle deductions from them" [. . .] we are told [elsewhere, by Huxley] that "Mathematics is that study which knows nothing of observation, nothing of experiment, nothing of induction, nothing of causation." I think no statement could have been made more opposite to the undoubted facts of the case, that mathematical analysis is continually invoking the aid of new principles, new ideas and new methods, not capable of being defined by any form of words, but springing directly from the inherent powers and activity of the human mind, and from continually renewed introspection of that inner world of thought of which the phenomena are as varied and require as close attention to discern as those of the outer physical world [...] that it is unceasingly calling forth the faculties of observation and comparison, that one of its principal weapons is induction, that it has frequent resource to experimental trial and verification, and that it affords a boundless scope for the exercise of the highest efforts of imagination and invention.'

Sylvester added in an Appendix that 'Induction and analogy are the special characteristics of modern mathematics.'

Ironically, T. H. Huxley 'made some of his most important discoveries as a result of an inherent aesthetic disposition to appreciate unities and symmetries in complex forms'. In fact, Huxley was a brilliant and pioneering student of biological structure – what a pity that he did not appreciate the role that structure was coming to play, even as he wrote, in mathematics – though, to be fair, Sylvester and other mathematicians of the period, did not realize it either.

An IQ test

'Some years ago a well-known public official in California quit his job and moved to Alabama, thereby – in the words of a local editorial – raising the average IQ in both states. That this should indeed be possible is apparent. By the same token, it is also possible, by a mere redistribution of the population of the United States, to raise the average IQ in all fifty states. Does this, then, imply that the average IQ of the entire country – being itself an average of the average IQs of its constituent states – can thus be raised?'

He's a poet, does he know it?

Dr William Whewell (1794–1866) was a distinguished philosopher of science, as well as being the author of *An Elementary Treatise on Mechanics* (1819) in which he inadvertently wrote this small poem-in-prose as a prose sentence. It is arranged here as a natural four-line verse:

> Hence no force however great,
> Can stretch a cord however fine,
> Into a horizontal line,
> Which is accurately straight.

It is not impossible that Whewell's natural feeling for language unconsciously contributed to this performance – he was the author of two published volumes of verse.

A popular innovation

'Lacroix was the enthusiastic promoter and populariser of *analytic* geometry, [what we now call coordinate geometry] in which more attention was paid to the algebraic analysis, and less to geometrical diagrams and traditional synthesis. Analytic geometry became quickly and immensely popular in the early decades of the nineteenth century.

F. W. Newman, brother of Cardinal Newman, obtained a Double First at Oxford in 1826. He is said to have been the first man who ever offered in the Schools the Higher Mathematics analytically treated. Only one of his examiners understood the subject; and he reported to his colleagues answers so brilliant that, besides awarding a First Class, they presented the candidate with finely bound copies of Laplace and Lagrange.'

Hadamard's failures

'Every scientist can probably record similar failures. In my own case, I have several times happened to overlook results which ought to have struck me blind, as being immediate consequences of other ones which I had obtained. Most of these failures proceed from the cause which we have just mentioned, viz., from attention too narrowly directed.

The first instance I remember in my life had to do with a formula which I obtained at the very beginning of my research work. I decided not to publish it and to wait till I could deduce some significant consequences from it. At that time, all my thoughts, like many other analysts', were concentrated on one question, the proof of the celebrated "Picard's theorem". Now, my formula most obviously gave one of the chief results which I found four years later by a much more complicated way: a thing which I was never aware of until years after, when Jensen published that formula and noted, as an evident consequence, the results which, happily for my self-esteem, I had obtained in the meanwhile. It is clear that, in 1888, I had thought too exclusively of Picard's theorem.

My next work was my thesis. Two theorems, important to the subject, were such obvious and immediate consequences of the ideas contained therein that, years later, other authors imputed them to me, and I was obliged to confess that, evident as they were, I had not perceived them.

Some years later, I was interested in generalizing to hyperspaces the classic notion of curvature of surfaces. I had to deal with Riemann's notion of curvature in hyperspaces, which is the generalization of the more elementary notion of the curvature of a surface in ordinary space. What interested me was to obtain that Riemann curvature is the curvature of a certain surface S, drawn in the considered hyperspace, the shape of S being chosen in order to reduce the curvature to a minimum. I succeeded in showing that the minimum thus obtained was precisely Riemann's expression; only, thinking of that question, I neglected to take into consideration the circumstances under which the minimum is reached, i.e., the proper way of constructing S in order to reach the minimum. Now, investigating that would have led me to the principle of the so-called "Absolute Differential Calculus", the discovery of which belongs to Ricci and Levi Civita.

Absolute differential calculus is closely connected with the theory of relativity; and in this connection, I must confess that, having observed that the equation of propagation of light is invariant under a set of transformations (what is now known as

Lorentz's group) by which space and time are combined together, I added that "such transformations are obviously devoid of physical meaning". Now, these transformations, supposedly without any physical meaning, are the base of Einstein's theory.

To continue about my failures . . .'

Omnipresence

'I have come to know that Geometry is at the very heart of feeling, and that each expression of feeling is made by a movement governed by Geometry. Geometry is everywhere in Nature. This is the Concert of Nature.'

Auguste Rodin (1840–1917)

Descartes' confidence

'Those long chains of perfectly simple and easy reasonings by means of which geometers are accustomed to carry out their most difficult demonstrations had led me to fancy that everything that can fall under human knowledge forms a similar sequence; and that so long as we avoid accepting as true what is not so, and always preserve the right order for deduction of one thing from another, there can be nothing too remote to be reached in the end, or too well hidden to be discovered. I had no great difficulty over looking for a starting-point. I knew already that I must start with the simplest objects, those most apt to be known; and seeing that, among all those who have so far sought for truth in the sciences, only mathematicians have been able to find some demonstrations, that is to say, some certain and self-evident reasonings, I had no doubt that I must start from the objects that they treated of. The only advantage I hoped for here was that I should habituate my mind to nourish itself on truths and not acquiesce in bad arguments.

I venture to say that the exact observance of the few rules I had chosen gave me such powers of unravelling all the problems covered by these two sciences that in the two or three months I spent in examining them I not only solved some that I had formerly considered very difficult, but was also in the end apparently able to determine by what means, and to what extent, a solution was possible, even in fields where I was still ignorant of one. To this end, I began with the simplest and most general problems; and every truth I discovered was a rule applicable towards further discoveries. My claim will not appear too conceited if you consider that, since there is only one truth in any matter, whoever discovers it knows as much about it as can be known. For instance, a child who has been taught arithmetic and does an addition according to the rules may be assured that he has discovered all that the human mind can discover as regards the sum he is considering.'

The grammar of art

'One can say that geometry is to the plastic arts what grammar is to the writer.'

Apollinaire

Mathematical medicine

'I once picked up a copy of Casey's *Sequel to Euclid* on a second-hand bookstall, and on the inside of the front cover was one of the blue-edged labels that doctors stick on medicine bottles with the inscription "Poison – to be taken three times a day".'

F. Bowman

Leonardo on Mathematics

'Let no man who is not a mathematician read the elements of my works.'

'Mechanics are the paradise of mathematics because here we come to the fruits of mathematics . . . There is no certainty in science where one of the mathematical sciences cannot be applied, or which are not in relation with these mathematics.'

'No knowledge can be certain, if it is not based upon mathematics or upon some other knowledge which is itself based upon the mathematical sciences. Instrumental, or mechanical science is the noblest and above all others, the most useful.'

'[Painting] can only be learnt by those to whom it has been granted by nature, unlike mathematics, in which the pupil acquires as much as the master reads to him.'

Sonya Kovalevskaya (1850–1891)

'One day Professor Weierstrass was rather surprised to see a young lady present herself before him, asking to be admitted as his pupil in mathematics. The Berlin University was, and still is, closed to women, but Sonia's ardent desire to be taught by the man who was generally acknowledged to be the father of modern mathematical analysis, made her apply to him for private lessons.

Professor Weierstrass felt a certain distrust in seeing this unknown female applicant; however, he promised to try her, and gave her some of the problems which he had set apart for the more advanced pupils in the seminary for mathematics. He felt convinced that she would not be able to solve them, and forgot all about her, the more so as her outward appearance on the first visit had left no impression at all upon his mind. She never dressed well, and on this occasion she wore a hat which hid her face completely, and made her look very old, so that Professor Weierstrass, as he told me himself, after having seen her for the first time, had neither the slightest idea of her age, nor of her unusually expressive eyes, which used to attract everybody at first sight. A week later she called again, and said that she had solved all the problems. He did not believe her, but asked her to sit down beside him, after which he began to examine her solutions one by one. To his great surprise everything was not only correct, but very acute and ingenious. Now in her eagerness she took off her hat and uncovered her short curly hair; she blushed at his praises, and the elderly professor felt something like fatherly tenderness towards this young woman, who possessed the divination of genius to a degree he had seldom found, even in his more advanced male pupils. And from that moment the great mathematician became her friend for life, the most faithful and helpful friend she could wish. In his family she was received as a daughter and sister.

It was her great object to find the logical connection between all manifestations of life, as for instance, between the laws of thought and the outward phenomena. She could not satisfy herself with seeing in part, and understanding in part; it was her delight to dream of a more perfect form of life, where, according to the apostle, "we shall see no longer in part, but face to face". To see the unity in the variety was the aim and end of all her philosophy and her poetry.

Has she reached this end now? Our thought cannot fathom this possibility, but our heart beats with a trembling hope which breaks the point of death's bitterness.

Besides, she had always wished to die young. Though hers seemed an inexhaustible well of life, ready for every new impression, open to every joy, great or small, in the innermost recess of her heart there was a thirst, which this life could never satisfy. As her mind craved absolute truth, absolute light, so her heart craved absolute love – a completeness which human life does not yield, and which her own character in particular rendered impossible. It was this discord that consumed her. If we start from her own belief in a fundamental connection between all phenomena of life, we see that she was bound to die, not because some strong and destructive microbes had settled in her lungs, or because the chances of her life had not brought her the happiness she desired, but because the necessary organic connection between her inward and outward life was missing; because there was no harmony between her thought and her feeling, her temperament and her character.'

Anna Carlotta Leffler

The craft of geometry

'. . . to talk freely with you about geometry is to me the very best intellectual exercise, but that at the same time I recognise it to be so useless that I can find little difference between a man who is only a geometrician and a clever craftsman. Although I call it the best craft in the world it is, after all, nothing else but a craft, and I have often said it is fine to try one's hand at it but not to devote all one's powers to it [. . .] it is quite possible I shall never think of it again . . .'

Pascal writing to Fermat, 10 August 1660

Pascal did indeed abandon mathematics, and died two years later.

Nicole Lepaut and the return of Halley's comet

'. . . by 1757, there were those who had become obsessed with the idea that the gravitational physics of Newton could actually be used to foretell the future. Among them was Alexis Clairaut, an eminent French mathematician, who had published his first paper at the age of thirteen. He made a last-minute decision to try to improve on Halley's tables on the orbit of the comet of 1682, and the predicted time of its return. It was imperative, of course, that the revised prediction appear before the comet did, "so that no one might doubt the agreement between the observation and the calculations". But the comet was fast approaching and the task was enormous, requiring meticulous calculation of the gravitational interactions of Jupiter, Saturn, the Earth and the comet over a period of 150 years. Clairaut claimed that he engaged the astronomer Joseph Jerome de Lalande to help him. To hear Lalande tell it, it was the other way around. But Clairaut made no public mention of the third member of the team, without whom – as Lalande was later to admit – they would never have dared to try to beat the comet; it was she who deserved most of the credit.

We can only imagine the forebearance that Madame Nicole-Reine Etable de la Briere Lepaute needed to get through her remarkable life. It was an age when upper-class women were valued for their appearance, their ability to oversee the running of the household, and their capacity for lively small talk. Madame Lepaute fulfilled these ideals, but she was also a first-rate mathematician. In this regard, she posed a problem for her colleagues that is made clear by Lalande's tribute to her in his *Astronomical Bibliography*. Writing soon after her death in 1788, Lalande goes to great lengths both to exalt and to belittle her. Yes, she was vital to their work on the comet, but she was not pretty enough. Yes, her tables of parallactic angles, and her accurate prediction for the whole of Europe of the annular eclipse of 1764, were important, but no, she herself was significant mainly in terms of her male relatives. He *will* give her this much: "Her calculations never got in the way of her household affairs; the ledgers were next to the astronomical tables." Clairaut suppressed knowledge of Mme Lepaute's contribution, Lalande wrote, "in order to accommodate a woman jealous of Mme Lepaute's merit, and who had pretensions but no knowledge whatsoever. She was able to have this injustice committed by a judicious but weak scientist whom she had subjugated." Scientific biographies were racier then.

But Lalande's devotion to her is clear:

Mme Lepaute was the only woman in France who acquired true insights into astronomy . . . She was so dear to me, that the day I walked in her funeral procession was the saddest I have spent since I learned of the death of my father . . . The times I spent near her and in the heart of her family are those I am most fond of, the memory of which, mixed with bitterness and pain, spreads some comfort over the last years of my life . . . Her portrait, which I still have before my eyes, is my consolation.

Lepaute must have had her hands full in 1757. With Clairaut and Lalande, she worked day and night, often through meals, for six months, in a desperate race with the comet. The enterprise was so taxing, Lalande later wrote, "that following this forced labour, I contracted an illness which would change my temperament for the rest of my life". Eventually, they discovered that the comet would be detained by Saturn's gravity by 100 days. Jupiter meant a delay of at least 518 days. In the course of their calculations, they found that Halley had made a set of compensating errors which cancelled each other out, and concluded that Halley's estimate of the time of the return was essentially correct.

In November of 1758, they predicted that the comet would achieve perihelion passage on 13 April 1759 and might be visible some months before. On Christmas night 1758, a German farmer – one Johann Palitzsch – became the first to know that the long-dead Edmond Halley had successfully employed Newton's Laws to foretell the future. The comet was punctual, and it came from just the sector of the sky that Halley had foretold. Palitzsch, an avid amateur astronomer, one of many to make a contribution to cometary astronomy, rushed to tell the world. Halley's prodigal comet had returned. It reached perihelion on 13 March 1759, within a month of the Clairaut–Lalande–Lepaute prediction. Science had succeeded where generations of mystics had failed. Newtonian prophecy had been fulfilled.

Many soon recognized what Halley and his French successors had accomplished. They had established a programme, a goal, an ideal for the future of all of science: "The regularity which astronomy shows us in the movements of the comets," Laplace concluded, "doubtless exists also in all phenomena." '

The dangers of pattern spotting

'His intuition worked in analogies [. . .] and [. . .] by empirical induction from particular numerical cases [. . .] The clear-cut idea of what is meant by a proof [. . .] he perhaps did not possess at all. If a significant piece of reasoning occurred somewhere, and the total mixture of evidence and intuition gave him certainty, he looked no further.'

Littlewood on Ramanujan

The result, however, was that Ramanujan's work, despite his marvellous intuition, occasionally contained false as well as true conclusions. Japanese traditional mathematics, called *wasan*, illustrated the same weakness.

'Imperfect induction [was] one of the characteristic features of the Japanese mathematics [. . .] mathematics as a branch of natural science [. . .] never thought of demonstration [. . .] In cases [like this] it was only customary to take the relation for granted from its existence in a few first instances [. . .] it resulted therefore that correct propositions intermingled with false results, which even the best mathematicians were not able to detect.'

Vision

To see a World in a Grain of Sand,
And a Heaven in a Wild Flower,
Hold Infinity in the Palm of your hand,
And Eternity in an hour.

William Blake (1757–1827)

Alonzo Church

'It cannot be a complete coincidence that several outstanding logicians of the twentieth century found shelter in asylums at some time in their lives: Cantor, Zermelo, Gödel, Peano, and Post are some. Alonzo Church was one of the saner among them, though in some ways his behavior must be classified as strange, even by mathematicians' standards.

He looked like a cross between a panda and a large owl. He spoke slowly in complete paragraphs which seemed to have been read out of a book, evenly and slowly enunciated, as by a talking machine. When interrupted, he would pause for an uncomfortably long period to recover the thread of the argument. He never made casual remarks: they did not belong in the baggage of formal logic. For example, he would not say: "It is raining." Such a statement, taken in isolation, makes no sense. (Whether it is actually raining or not does not matter; what matters is consistency.) He would say instead: "I must postpone my departure for Nassau Street, inasmuch as it is raining, a fact which I can verify by looking out the window." (These were not his exact words.) Gilbert Ryle has criticized philosophers for testing their theories of language with examples which are never used in ordinary speech. Church's discourse was precisely one such example.

He had unusual working habits. He could be seen in a corridor in Fine Hall at any time of day or night, rather like the Phantom of the Opera. Once, on Christmas day, I decided to go to the Fine Hall library (which was always open) to look up something. I met Church on the stairs. He greeted me without surprise.

Every lecture began with a ten-minute ceremony of erasing the blackboard until it was absolutely spotless. We tried to save him the effort by erasing the board before his arrival, but to no avail. The ritual could not be disposed of; often it required water, soap, and brush, and was followed by another ten minutes of total silence while the blackboard was drying. Perhaps he was preparing the lecture while erasing; I don't think so. His lectures hardly needed any preparation. They were a literal repetition of the typewritten text he had written over a period of twenty years, a copy of which was to be found upstairs in the Fine Hall library. (The manuscript's pages had yellowed with the years, and smelled foul.

Church's definitive treatise was not published for another five years.) Occasionally, one of the sentences spoken in class would be at variance with the text upstairs, and he would warn us in advance of the discrepancy between oral and written presentation. For greater precision, everything he said (except some fascinating side excursions which he invariably prefixed by a sentence like: "I will now interrupt and make a metamathematical [sic] remark") was carefully written down on the blackboard, in large English style handwriting, like that of a grade-school teacher, complete with punctuation and paragraphs. Occasionally, he carelessly skipped a letter in a word. At first we pointed out these oversights, but we quickly learned that they would create a slight panic, so we kept our mouths shut. Once he had to use a variant of a previously proved theorem, which differed only by a change of notation. After a moment of silence, he turned to the class and said: "I could simply say 'likewise', but I'd better prove it again."

It may be asked why anyone would bother to sit in a lecture which was the literal repetition of an available text. Such a question would betray an oversimplified view of what goes on in a classroom. What one really learns in class is what one does not know at the time one is learning. The person lecturing to us was logic incarnate. His pauses, hesitations, emphases, his betrayals of emotion (however rare), and sundry other nonverbal phenomena taught us a lot more logic than any written text could. We learned to think in unison with him as he spoke, as if following the demonstration of a calisthenics instructor. Church's course permanently improved the rigor of our reasoning.'

Gian-Carlo Rota

Mondrian versus the computer

Modern artists have often been attracted by chance events or surprise juxtapositions. Jean Arp painted 'Rectangles arranged according to the Laws of Chance' as early as 1916. The elements in some of Mondrian's paintings might *seem* to be placed by chance – but are they?

'A digital computer and microfilm plotter produced a semi-random picture similar to Mondrian's "Composition with lines"

(1917). Only 28 out of 100 American viewers could tell which was the genuine article, and 59 actually preferred the computer-generated picture.'

The conviction of induction

How confident can a mathematician be when confronted by an 'obvious' pattern? It depends on the context. Many mathematicians would be easily convinced by an induction in geometry which consisted of drawing a physical figure, making an observation about it, and then repeating that observation on a few variants of the diagram – where 'few' might mean very few indeed. Indeed, such inductions have been the source of many geometrical theorems.

Patterns in numbers can also be very persuasive. Euler wrote of one series: 'For each of us can convince himself of this truth by performing the multiplication as far as he may wish; and it seems impossible that the law which has been discovered to hold for 20 terms, for example, would not be observed in the terms that follow.'

Major MacMahon remarked of one series: 'This most remarkable theorem has been verified as far as the coefficient of x^{89} by actual expansion so that there is practically no reason to doubt its truth; but it has not yet been established.'

On the other hand, Richard Guy has provided in two papers eighty examples of plausible inductions for readers to exercise their judgement on – many do not work. Here is a famous example, used frequently by schoolteachers to correct the naïve expectations of their pupils:

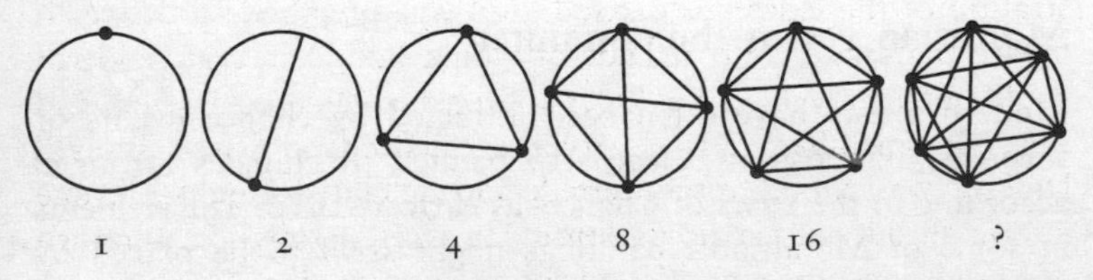

The numbers under each circle show the maximum number of parts into which it can be divided by the lines joining 1, 2, 3, 4, etc. dots on its circumference. The 'obvious' expected number for the final diagram is 32, but the correct number is actually 31.

Dr Johnson's advice

'Johnson said that a man had repeatedly called on him, refusing to leave his name. When they did meet, the man confessed that he was oppressed with scruples of conscience [. . .] it occurred to him that the man might be mad, so he asked him how his day was spent, and found he had five hours unemployed. "Five hours of the twenty-four are enough for a man to go mad in; so I would advise you, sir, to study algebra, if you are not adept already in it; your head would get less muddy" [. . .] When Johnson felt his own fancy disordered he fled to arithmetic, but probably thought this man had enough arithmetic in his counting-house.'

James Boswell

Diderot at Court

'In 1773 Diderot spent some months at the court of St Petersburg at the invitation of the Russian empress, Catherine the Great. He passed much of his time spreading his gospel of atheism and materialism among the courtiers, until it was suggested to the empress that it would be desirable to muzzle her guest. Reluctant to take direct action, Catherine requested the aid of another savant, the Swiss mathematician Leonhard Euler, a devout Christian. As Diderot was almost entirely ignorant of mathematics, a plot was hatched to exploit this weakness. He was informed that a learned mathematician had developed an algebraical demonstration of the existence of God, and was prepared to deliver it before the entire court if Diderot would like to hear it. Diderot could not very well refuse. Euler approached Diderot, bowed, and said very solemnly, "Sir, $(a + b^n)/n = x$, hence God exists. Reply!" Diderot was totally disconcerted, and delighted laughter broke out on all sides at his discomfiture. He asked permission to return to France, and the empress graciously consented.'

The wrong brother

C. G. J. Jacobi (1804–1851) was the brilliant rival of the equally brilliant and short-lived Abel (see page 149). Like all mathematicians he had a proper sense of his own worth.

'Jacobi's brother, M. H. Jacobi, had a prodigious contemporary reputation as the founder of the fashionable "science" of galvanoplasty. The professor of mathematics was constantly being mistaken for M. H. Jacobi or even congratulated on having such a famous sibling. Conscious of the lasting value of his own work, C.G.J. found this tiresome. When a lady complimented him on having such a distinguished brother, he retorted, "Pardon me, madame, but *I* am my brother." '

The USSR against G. H. Hardy

'This story started at the end of the 1950s and has now finished. I was a student in the Mathematics Department of Leningrad (now St Petersburg) University. One day our lecturer on the History of the Mathematics called me after his lecture and asked: "Do you know the name Hardy?" "Of course, I have heard of him," I answered. "He was one of the most prominent British mathematicians." "Well, he continued, Hardy died in 1948, and some years ago I found the issue of 'Nature' with Hardy's obituary. I was very surprised that one paragraph in that short obituary was cut out! Hardy, was a pure mathematician, and far from politics. I tried to check in another library the same result: that paragraph was cut out. What does it mean? You are young, the lecturer told me, maybe sometime you will find out what was in that paragraph," he finished.

I immediately ran to the main library, the State Public Library – and found that issue of "Nature". My professor was quite right: there was a hole instead of the small paragraph in the end of the obituary. I tried to imagine what kind of "antisoviet propaganda" could be in that article.

Here I want to explain that in that time (during Stalin's era and later till the 80s) Soviet censorship was incredibly savage. Any mention of Soviet history, Soviet leaders, Soviet affairs and realities had to be either in support or had to be cut out. Only a few scientific foreign journals were accessible and almost no popular magazines could be found in public libraries. There were no private subscriptions and even those magazines which were allowed had been strictly censored and a lot of articles were pulled out or had gaps as in our case. Some of the journals you could read only by special permission of the administration in so called special rooms, but such permission could not be obtained by everybody. The censors tried to preserve the "innocence" of the consciousness of the people!

That lecturer died many years ago and I had forgotten the story. For a long time – up to the 1990s – I had no permission to go abroad. Apparently the KGB did not trust me as that professor had. During my first visit to the UK as a Kapitza Fellow I recalled the story and as more than 30 years had elapsed, went to the library of the University for the same reason: to find the issue of "Nature" No. 4099, May 22, 1948, p. 798, and to read the obituary of Hardy, perhaps, without gaps. With great impatience I looked for the paragraph. Now I give you the opportunity to read that paragraph in its entirety:

> Hardy had one ruling passion – mathematics. Apart from that his main interest was in ball-games, of which he was a skilled player and an expert critic. An illustration of some of his interests and antipathies is given by this list of "six New-Year wishes" which he sent on a postcard to a friend (in the 1920s):
>
> (1) prove the Riemann hypothesis;
> (2) make 211 not out in the fourth innings of the last Test Match at the Oval;
> (3) find an argument for the non-existence of God which shall convince the general public;
> (4) be the first man at the top of Mount Everest;
> (5) be proclaimed the first president of the USSR of Great Britain and Germany;
> (6) murder Mussolini.

What can we say? How stupid and malicious were the ideologists in communist Russia, who had considered even a professor's jokes as unacceptable for the poor Soviet readers! How many stories of such a type one can find – this is only one of millions of examples. And even this funny story is rather sad, isn't it?'

A. M. Vershik

The heavenly spheres

Kepler believed according to the Pythagorean theory of the music of the spheres that each heavenly body produced a characteristic sound. The Earth, he said, gave out two notes, *mi* and *fa*, standing for misery (*miseria*) and famine (*fames*).

Poetry and Mathematics

'One of the most evocative things that Paul Valéry ever said was that he wished his poems to have "the solidity of certain pages of algebra". As a poet who has been an amateur mathematician and who was, once upon a time, for three years, a teacher of advanced high-school mathematics and calculus, I immediately recognize what Valéry is talking about. Passages of algebra, indented and breaking free from prose text, visually resemble passages of poetry, as if, out of the plodding welter of expository prose and fastidious explanation, some higher, more intense language had suddenly burst forth like a pure aria issuing from recitative – for example, the elementary proof by mathematical induction that, starting with the integer 1, the sum of the first n odd integers is the square of n:

$$\sum_{p=1}^{n} (2p-1) = n^2 \rightarrow \sum_{p=1}^{n+1} (2p-1) =$$

$$\sum_{p=1}^{n} (2p-1) + [2(n+1)-1] = n^2 + 2n + 1 = (n+1)^2.$$

When I was much younger, such flourishes of nomenclature seemed to me glamorous, abstruse, runic, and eternal, everything which, as an undergraduate poet laboring at wheezing translations from French symbolist poems, I knew that a poem should be – a hermitage of pure mind to which one could retreat from the ugly, inconvenient, and merely provisional aspects of sublunary life.'

Jonathan Holden

If only . . .

Roger Cotes (1682–1716) was dead before the end of his thirty-fourth year, prompting Newton to say that, 'If he had lived, we might have known something.'

When Abel (1802–1829) died, of tuberculosis, aged 27, Crelle, in Crelle's journal, claimed that if Abel had been a contemporary of Newton, Newton would have paid him the same compliment.

Galois, the schoolboy

Evariste Galois (1811–1832) died in a duel at the age of twenty, having already done brilliant mathematics. These extracts from his school reports speak for themselves.

1826–1827

This pupil, though a little queer in his manners, is very gentle and seems filled with innocence and good qualities . . . He never knows a lesson badly: either he has not learned it at all or he knows it well . . .

A little later:

This pupil, except for the last fortnight during which he has worked a little, has done his classwork only from fear of punishment . . . His ambition, his originality – often affected – the queerness of his character keep him aloof from his companions.

1827–1828

Conduct rather good. A few thoughtless acts. Character of which I do not flatter myself I understand every trait; but I see a great deal of self-esteem dominating. I do not think he has any vicious inclination. His ability seems to me to be entirely beyond the average, with regard as much to literary studies as to mathematics . . . He does not seem to lack religious feeling. His health is good but delicate.

Another professor says:

His facility, in which one is supposed to believe but of which I have not yet witnessed a single proof, will lead him nowhere: there is no trace in his tasks of anything but of queerness and negligence.

Another still:

Always busy with things which are not his business. Goes down every day.

Same year, but a little later:

> Very bad conduct. Character rather secretive. Tries to be original . . . Does absolutely nothing for the class. The furor of mathematics possesses him . . . He is losing his time here and does nothing but torment his masters and get himself harassed with punishments. He does not lack religious feeling; his health seems weak.

Later still:

> Bad conduct, character difficult to define. Aims at originality. His talents are very distinguished; he might have done very well in "Rhétorique" if he had been willing to work, but swayed by his passion for mathematics, he has entirely neglected everything else. Hence he has made no progress whatever . . . Seems to affect to do something different from what he should do. It is possibly to this purpose that he chatters so much. He protests against silence.

Mancala, a mathematical game

The game known as *mancala* or *soro* or *igisoro* or *ikiokoto* or *ayo* or *wari* or *owari*, among dozens of other names, is played in scores of variant forms throughout Africa, parts of Asia, and the West Indies and South America where it was introduced by slaves.

Most boards have two rows of shallow holes, one row belonging to each player. Initially, each hole contains a specified number of seeds, or shells. A turn consists of scooping the seeds from one hole and distributing them, one each, into subsequent holes, in the traditional direction. If a player's last seed falls into an enemy hole containing one or two counters, he captures the contents of that hole.

'Several decades ago there lived in the western part of Tanganyika (now Tanzania) a youngster named Kamberage. He delighted in playing *soro* with his father's friend, Chief Ihunyo, and no wonder, since the boy always came out the winner in this complicated African board game. Much impressed by Kamberage's skill, the older man advised his friend, Chief Nyerere, to send his son to school. Nyerere accepted the suggestion, and soon Kamberage was enrolled in a Native Authority boarding school in Musoma. In no time at all he was at the head of his class.

No doubt Julius Nyerere, the president of Tanzania, would have attended school even if he had not excelled in *soro*.

In past centuries each new Kabaka of Buganda, upon his accession, was required to perform the ritual of *okweso*. The new king picked brown seeds from a certain tree, and these were later used in the *omweso* board game kept in the royal hall. Here the Prime Minister played *omweso* while he decided legal cases. The significance of this ritual is that the Kabaka shall not be outwitted by his people. If they should try to trick him, he would overcome them by his strategy, just as an *omweso* expert defeats his opponent by a clever move.

Although the Kabaka of Buganda is said to have played the game with his wives and sisters, it is rarely played by men and women together. Apparently no man's reputation could bear the ridicule that would be heaped upon him if he were defeated by a woman! Ganda women outside of the royal court were not allowed to play for fear their crops would fail. Young girls were warned that their breasts would not develop, and no men would marry them. Thus the men were assured that the game would not distract the women and girls from their assigned chores in the fields and the home.

Chief Ayorinde informed me about the customs and etiquette of the game among the Yoruba. It may be played indoors or out, usually after work, but rarely at night. Some chiefs and elders are willing to spend any amount of money to invite players of high repute to their homes for a few matches. He remarks about two methods of play; it may be played just for fun and relaxation, involving little drain on the intellect, or it may be a game of great skill, "as interesting to watch as watching a man who knows his trigonometry inside out and is proving a problem with ease".

Every observer has remarked upon the speed and skill exhibited by the contestants. One is not allowed to pick up the contents of a hole to count. The good player remembers the number of seeds, not only in his own holes, but in those of his opponent . . .'

With a little help from my friends

'I am a mathematician to this extent: I can follow triple integrals if they are done slowly on a large blackboard by a personal friend.'

J. W. McReynolds

One pea = the sun

The Banach–Tarski paradox says that it is possible to cut up a sphere into a finite number of pieces which can then be reassembled to form two spheres, each the same size as the original sphere. Moreover, not more than five pieces are needed for this trick.

Indeed, it is possible to cut and reassemble a given sphere to make any finite number of copies of itself, or indeed an enumerable infinity of such copies. Or you could cut up a pea and reassemble the pieces to make a ball the size of the sun.

It is called a paradox because even professional mathematicians find it extremely counter-intuitive, though in fact there is nothing paradoxical about it, provided you realize – for a start – that the pieces are not of any simple shape. In particular, they do *not* each have a volume which can be measured in the usual way.

Look inside the box

'I [planned to] write so that the learner may always see the inner ground of the things he learns, even so that the source of the invention may appear, and therefore in such a way that the learner may understand everything as if he had invented it himself.'

Leibniz

The value of experience

'It may be said that the fact makes a stronger impression on the boy through the medium of his sight, that he believes it the more confidently. I say that this ought not to be the case. If he does not believe the statements of his tutor, probably a clergyman of mature knowledge, recognised ability, and blameless character – his suspicion is irrational, and manifests a want of the power of appreciating evidence, a want fatal to his success in that branch of science which he is supposed to be cultivating.'

Isaac Todhunter

The Dining Philosophers Problem

Traditional puzzles appeal partly because their setting makes tricky ideas easier to think about. This is as true today, as this explanation of an important problem in computing illustrates:

'We have a table, around which are seated *N* philosophers. In the center there is a large plate containing an unlimited amount of spaghetti. Half-way between each pair of adjacent philosophers there is a single fork (see Figure). Now, since no one, not even a

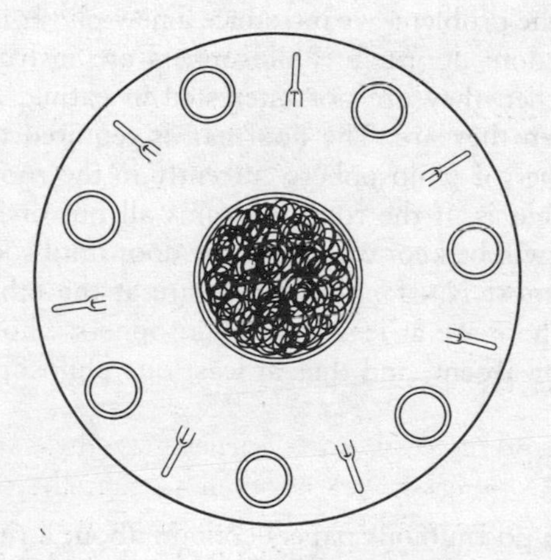

philosopher, can eat spaghetti with one fork, there is a problem. The desired life cycle of a philosopher consists of carrying out his private activities (e.g., thinking, and then writing up the results for publication), becoming hungry and trying to eat, eating, and then back to the private activities, *ad infinitum*. (A related problem involves Chinese philosophers, with rice and chopsticks replacing spaghetti and forks.)

How should the philosophers go about their rituals without starving? We can instruct them to simply lift up the forks on either side when hungry, and eat, laying them down again when finished.

This solution will not work, since one or both of the forks may be taken by neighboring philosophers at the time. Also, two adjacent philosophers may try to pick up the same fork at the same time. Using forks that are beyond a philosopher's reach is forbidden. Here, eating can be considered a critical section of sorts, as two adjacent philosophers cannot eat simultaneously, and the forks are crucial resources of sorts, as they cannot be used by two philosophers simultaneously.

This problem typifies many real-world situations, such as computer operating systems, for example, in which many processors compete for certain shared resources. The interconnection pattern of such systems is usually quite sparse (not all resources are accessible to each processor) and the number of resources is too small for everyone to be happy together.

To solve the problem, we introduce a new player into the game, the dining room doorman. Philosophers are instructed to leave the room when they are not interested in eating, and to try to re-enter when they are. The doorman is required to keep count of the number of philosophers currently in the room, limiting it to $N - 1$. That is, if the room contains all but one philosopher, the last one will be kept waiting at the door until someone leaves. Now, if at most $N - 1$ philosophers are at the table at any one time, then there are at least two philosophers who have at least one neighbor absent, and thus at least one philosopher can eat.'

Full Marx

Karl Marx's posthumous papers contain about a thousand pages of mathematical manuscripts, mainly dealing with the foundations of the calculus.

The basic facts

P. G. Lejeune-Dirichlet (1805–1859) invented, among other achievements, Dirichlet's function, $f(x)$, which has one value when x is rational and a different value when it is irrational.

'Dirichlet was opposed to writing letters; many of his friends had in the course of their entire lives received no communication from him. However, when his first child was born he broke his silence, he wired his father-in-law: "2 + 1 = 3".'

The deaths of two Jewish mathematicians

Mark Kac (1914–) was Jewish, born in Poland where his family remained at the start of the Second World War, in the town of Krzemieniec. They perished there in mass executions during 1942–3. Kac took a position in the United States in 1938, little realizing what lay ahead. Many mathematicians were less fortunate. This is Kac's account of the death of Wladek Hetper, his closest friend in the years preceding the outbreak of war. It is followed by an abbreviated account of the death of Dr Ludwig Berwald, who at the outbreak of war was Professor of Mathematics at the German University in Prague.

'In less than a year the world exploded and much of my part of it was consumed by flames. Millions, including my parents and my brother, were murdered by the Germans and many disappeared without a trace in the vastness of the Soviet Union. Wladek Hetper was one of them. In 1939, before the outbreak of the war, he received his *veniam legendi*. When hostilities started, as a reserve officer he was called to active duty and sent to the Eastern Front. I heard nothing more of him until, long after the war, an article in a London-based Polish literary journal printed excerpts from the diary of a Feliks Lachman who had spent some time with Wladek in a Soviet prisoners' camp in 1940. He wrote a short, moving account of their friendship, which was interrupted by Lachman's sentencing and deportation to Siberia. Wladek, according to this account, was in poor health, which, considering what a superb physical specimen he had been, leads me to conjecture that he was suffering from malnutrition. On a visit to London I contacted Lachman and some time later (April 1974) he sent me an excerpt from his diary:

1940, December, Starobielsk The Fourth Dimension

My acquaintance with Wladek the mathematician lasted for not longer than five weeks. Like the homeless swallow in the Norse saga, he flew into my life out of darkness, and into darkness he went away. I never met him again, neither do I know where his bones lie now. He was thirty-one when I met him, a pessimist, though a deeply religious man. He played chess marvelously and could solve difficult mathematical problems in his memory. We spent many a long and thrilling evening discussing Bertrand Russell, the principles of topology and entertaining one another with physical-mathematical puzzles and intelligence tests. We succeeded in reconstructing Cardano's solution of the cubic equation. And then, head first, we plunged into multi-dimensional worlds.

This was the last glimpse I had of my friend in a cruelly distorted mirror image of our times together in Lwów. We, too, had talked of Russell and topology and had entertained each other with all sorts of puzzles. And cubic equations! Another shadow of the past? I'll never know.'

On 22 October 1941, at the collecting point located next to the Messepalast in Prague, the third transport of Jews to be deported to the Ghetto in Lodz, Poland, was assembled by order of the German Secret Police.

'The Berwalds were included in this transport. Their registration numbers were 2793/816 and 817. The preceding day, Berwald had distributed his last mathematical manuscripts. On October 22nd his scientific work came to a close. It would still have been possible, through a medical certificate, to withdraw Nos. 2793/816 and 817 from Transport C. However, these numbers wanted no part of such a maneuver and chose to put their lot in with the one thousand other people of this transport, thereby assuring themselves of certain death. When Transport C arrived in Lodz, the "native" Jews there told the newcomers that they had been in the Ghetto already for two years. In fact, the Ghetto of Lodz had been created shortly after the German invasion in 1939. The Berwalds were placed in an incompleted one-story schoolhouse whose raw bricks were not yet covered, in the former Marenšinska then called Siegfriedstrasse 48. There were no beds, straw pallets, or even piles of straw for the occupants to lie on. People simply lay down next to each other on the bare floor. Some were also placed in the attic. The typical diet consisted of ¼ kilogram of bread daily; in addition to this they received black "coffee" twice daily, and, once a day, a miserable soup. The Jews were allowed to move about freely within the Ghetto. The younger people were drafted to perform labor. One group of young people were later transported to Posen where they died of typhus. The mass transportation of Jews to extermination camps from Lodz began in 1942. In May 1942, 12,000 elderly and unemployable people were transported to Majdanek; 20,000 in the Fall of 1942. There they were murdered en masse by methods even more brutal than those used in Auschwitz. Since Berwald did not work he most probably would have been murdered in Majdanek if he had lived until May or certainly by the Fall of 1942. However, Mrs Berwald died on March 27, 1942, and Professor Ludwig Berwald died a few weeks later on April 20th.'

Hardy visits Ramanujan

'Hardy used to visit him, as he lay dying in hospital at Putney. It was on one of those visits that there happened the incident of the taxi-cab number. Hardy had gone out to Putney by taxi, as usual his chosen method of conveyance. He went into the room where Ramanujan was lying. Hardy, always inept about introducing a conversation, said, probably without a greeting, and certainly as his first remark: "The number of my taxi-cab was 1729. It seemed to me rather a dull number." To which Ramanujan replied: "No, Hardy! No, Hardy! It is a very interesting number. It is the smallest number expressible as the sum of two cubes in two different ways." '

C. P. Snow

Hilbert's breakfast

'Hilbert . . . was perhaps the most absentminded man who ever lived. He was a great friend of the physicist James Franck. One day when Hilbert was walking in the street he met James Franck and he said, "James, is your wife as mean as mine?" Well, Franck was rather taken aback by this statement and didn't know quite what to say, and he said, "Well, what has your wife done?" And Hilbert said, "It was only this morning that I discovered quite by accident that my wife does not give me an egg for breakfast. Heaven knows how long that has been going on." '

Fabre, spiders, and geometry

Jean Henri Fabre (1823–1915) was a brilliant entomologist who demonstrated that insects behaved by instinct and not by reasoning comparable to more evolved species. Curiously, although Darwin admired his work, and cited him in *The Origin of Species*, Fabre never accepted Darwin's theory of evolution.

Fabre referred to himself as 'a surveyor of spider's webs' and in a geometrical appendix to his *The Life of the Spider*, he combined all his marvellous talents of observation and analysis:

'Let us direct our attention to the nets of the Epeirae . . . We shall first observe that the radii are equally spaced; the angles formed by each consecutive part are of perceptibly equal value; and this in spite of their number, which in the case of the Silky Epeira exceeds two score. We know not by what strange means the spider attains her ends and divides the area wherein the web is to be warped into a large number of equal sectors, a number which is almost invariable in the work of each species . . .

We shall also notice that, in each sector, the various chords, the elements of the spiral windings, are parallel to one another and gradually draw closer together as they near the center. With the two radiating lines that frame them they form obtuse angles on one side and acute angles on the other; and these angles remain constant in the same sector, because the chords are parallel.

There is more than this: these same angles, the obtuse as well as the acute, do not alter in value, from one sector to another, at any rate so far as the conscientious eye can judge. Taken as a whole, therefore, the rope-latticed edifice consists of a series of crossbars intersecting the several radiating lines obliquely at angles of equal value.

By this characteristic we recognize the "logarithmic spiral". Geometricians give this name to the curve which intersects obliquely, at angles of unvarying value, all the straight lines or "radii vectores" radiating from a center called the "pole". The Epeira's construction, therefore, is a series of chords joining the intersections of a logarithmic spiral with a series of radii. It would become merged in this spiral if the number of radii were infinite, for this would reduce the length of the rectilinear elements indefinitely and change this polygonal line into a curve . . . The Epeira winds nearer and nearer around her pole so far as her equipment, which like our own, is defective, will allow her. One would believe her to be thoroughly versed in the laws of the spiral.'

Immersion

According to Littlewood, you need to know a problem as well as you know your own tongue going round your mouth, which invites the question: how well do most people know their own tongues? Anyway, here is Atiyah's view of the same subject:

'It is hard to communicate understanding because that is something you get by living with a problem for a long time. You study it, perhaps for years, you get the feel of it and it is in your bones. You can't convey that to anybody else. Having studied the problem for five years you may be able to present it in such a way that it would take somebody else less time to get to that point than it took you. But if they haven't struggled with the problem and seen all the pitfalls, then they haven't really understood it.'

Beauty versus effectiveness

'When a sudden illumination invades the mathematician's mind . . . it sometimes happens . . . that it will not stand the test of verification . . . it is to be observed almost always that this false idea, if it had been correct, would have flattered our natural instincts for mathematical elegance.'

Poincaré (1854–1912)

Scientists can also find beauty too attractive not to be true. James Watson having thought of the idea of helical symmetry in DNA molecules, found it 'too pretty not to be true'.

Geometry rules, OK

'I have no fault with those who teach geometry. That science is the only one which has not produced sects; it is founded on analysis and on synthesis and on the calculus; it does not occupy itself with probable truth; moreover it has the same method in every country.'

Frederick the Great (1712–1786)

Lovers

'Mathematicians are like lovers . . . Grant a mathematician the least principle, and he will draw from it a consequence which you must also grant him, and from this consequence another.'

Fontenelle (1657–1757)

A mathematical bent

'Mark all mathematical heads which be wholly and only bent on these sciences, how solitary they be themselves, how unfit to live with others, how unapt to serve the world.'

Roger Ascham (1515–1568)

Odd Data

If Western Europe continues to subside at its present rate of about 2.5 cm every ten years, the top of the Eiffel Tower will become a small island in the Atlantic in roughly 120,000 years time.

If all the salt dissolved in the seas of the world were crystallized, a layer 45 metres thick could be spread out over the entire land surface of the globe. This volume of salt would be fifteen times greater than the European land mass above sea level.

If, after the Flood (supposed to have taken place around 3000 BC) Noah had set himself to string electrons on a thread at the rate of one a second for eight hours a day, the chain so formed would today still be only two tenths of a millimetre long! Composed of hydrogen atoms, the chain would be something like 25 metres by now.

At every breath, each of us is likely to inhale some 50 million molecules of air exhaled at some time by Leonardo da Vinci or, for that matter, by any other person who lived for 65 years and whose breath has been well distributed throughout the atmosphere. Of Julius Caesar's *dying* breath, you can expect to inhale about six molecules.

According to the theory of thermodynamics, there is a non-zero possibility that a cubic centimetre of gas, in a sealed vessel, would spontaneously contract to half its volume, leaving the other half of the vessel empty, and then revert to its previous state. However, the fluctuation would take only about 10^{-4} seconds, and would be expected to occur only about once in every $10^{(10^{19})}$ years.

Genius and Insanity

'It is said that mathematicians are exempt from psychical derangements, but this is not true; it is sufficient to recall not only Newton . . . but the two famous distractions of Archimedes, the hallucination of Pascal, and the vagaries of the mathematician Codazzi . . . Codazzi was sub-microcephalic, oxycephalic, alcoholic, sordidly avaricious; to effective insensibility he added vanity so great that while still young he set apart a sum for his own funeral monument, and refused the least help to his starving parents; he admitted no discussion of his judgment even if it only concerned the cut of a coat; and he had taken it into his head that he could compose melodic music with the help of the calculus.

All mathematicians admire the great geometer Bolyai, whose eccentricities were of an insane character; thus he provoked thirteen officials to duels and fought with them, and between each duel he played the violin, the only piece of furniture in his house; when pensioned he printed his own funeral card with a blank date, and constructed his own coffin – a vagary which I have found in two other mathematicians who died in recent years. Six years later he had a similar funeral card printed, to substitute for the other which he had not been able to use. He imposed on his heir the obligation to plant on his grave an apple-tree, in remembrance of Eve, of Paris, and of Newton. Such was the great reformer of Euclid.'

Cesare Lombroso

The mind of a mnemonist

Sylvester was once approached by one of his research students who proposed to use a certain result in his research. Sylvester objected that the claimed theorem could not possibly be true, at which point his student tactfully explained to him that he, Professor Sylvester, had proved it himself, many years previously.

Euler, Cayley, Ramanujan, MacMahon – many mathematicians have had superb memories, but a good memory is not necessary to do mathematics. Perhaps it depends on *how* the mathematician thinks.

Aleksandr Luria was a very distinguished Russian psychologist, who wrote a book about a subject, 'S', who possessed a phenomenal, and highly visual, memory. In the following passages, as a result of his peculiar mode of thinking, he gets confused over 'nothing'.

'In order for me to grasp the meaning of a thing, I have to see it . . . Take the word *nothing*. I read it and thought it must be very profound. I thought it would be better to call *nothing* something . . . for I see this *nothing* and it is something . . . If I'm to understand any meaning that is fairly deep, I have to get an image of it right away. So I turned to my wife and asked her what *nothing* meant. But it was so clear to her that she simply said: "*Nothing* means there is nothing." I understood it differently. I saw this *nothing* and felt she must be wrong. The logic we use, for example. It's been worked out on the basis of years of experience. I can see how it has developed, and what it means to me is that one has to rely on his own sensations of things. If *nothing* can appear to a person, that means it is something. That's where the trouble comes in . . .'

The needs of the architect

The Roman architect Vitruvius, author of *De Architectura,* or, *The Ten Books of Architecture*, flourished, according to the best estimates, in the first century BC. In it he covered every subject from the education required of the architect, (the subject of the following quotation,) to the advantages to health of avoiding winds in dwellings, the acoustics of temples, aqueducts, wells and cisterns, sundials and water clocks, mechanical aids, including military catapults and siege machines, the best proportions for rooms, as well as the practical skills required of the architect and the principles of construction from the foundations upwards.

The *ballistae*, *catapultae* and *scorpiones* referred to are military machines. Typically, architecture, like mathematics and engineering, has always contributed to wars and their successful prosecution.

'Let him be educated, skilful with the pencil, instructed in geometry, know much history, have followed the philosophers with attention, understand music, have some knowledge of medicine, know the opinions of jurists, and be acquainted with astronomy and the theory of the heavens . . .

Geometry . . . is of much assistance in architecture, and in particular it teaches us the use of the rule and compasses . . . By means of optics, again, the light in buildings can be drawn from fixed quarters of the sky . . . by arithmetic the total cost of the buildings is calculated and measurements computed, but difficult questions involving symmetry are solved by means of geometrical theories and methods . . .

As for philosophy, it makes an architect high-minded and not self-assuming, but rather renders him courteous, just, and honest without avariciousness . . . Let him not be grasping or have his mind preoccupied with the idea of receiving perquisites, but let him with dignity keep up his position by cherishing a good reputation . . .

Music, also, the architect ought to understand so that he may have knowledge of the canonical and mathematical theory, and besides be able to tune *ballistae*, *catapultae* and *scorpiones* to the proper key. For to the right and left in the beams are the holes in the frames through which the strings of twisted sinew are stretched by means of windlasses and bars, and these strings must not be clamped and made fast until they give the correct note to the ear of a skilled workman . . .'

Wall wisdom

'The law of the excluded middle either rules or does not rule, OK?'

Graffiti

How many computers?

'That arithmetic is the basest of all mental activities is proved by the fact that it is the only one that can be accomplished by a machine.'

Schopenhauer (1788–1860)

In 1946 the National Physical Laboratory approved plans for ACE = Automatic Computing Engine, and suggested that 'one machine would suffice [for] the whole country'.

Thomas Watson of IBM had once predicted that only a handful of computers would ever be needed in the whole world. As late as 1954 when the IBM 650 was introduced, the company estimated total sales of 250. More than 1,500 were actually sold.

These bizarre failures of prophecy appear less absurd when some of the acronyms which served as titles for the early computers are unwound. For example, ENIAC (1943–6) stood for Electronic Numerical Integrator and Computer, and MANIAC (1952) was the Mathematical Analyzer, Numerator, Integrator And Computer. The message from these acronyms is that the early designers had their sights fixed as much on using digital computers to speed up calculations already done by mechanical calculating machines, for example to calculate tables faster – the ENIAC reduced the computing time for a firing table from one month to one day – as on imagining what new uses they might serve. Today, millions of computer users are not calculating anything – I am an example as I use an IBM clone to word-process this book.

Taking it nice and easy

Russell claimed that mathematics can be reduced to logic, and wrote, with his friend and one-time teacher, A. N. Whitehead, the three-volume *Principia Mathematica* to prove it. However, he did not make the mistake of claiming that this reduction is easy. In the middle of page 379 of the first volume, appears paragraph 54.43. Note the final sentence!

‘*4·43. $\vdash :. \alpha, \beta \varepsilon 1 . \supset : \alpha \cap \beta = \Lambda . \equiv . \alpha \cup \beta \varepsilon 2$

Dem.

$\vdash . {*}54\cdot26 . \supset \vdash :. \alpha = \iota\text{‘}x . \beta = \iota\text{‘}y . \supset : \alpha \cup \beta \varepsilon 2 . \equiv . x \neq y .$

[*51·231] $\equiv . \iota\text{‘}x \cap \iota\text{‘}y = \Lambda .$

[*13·12] $\equiv . \alpha \cap \beta = \Lambda$ (1)

$\vdash . (1) . {*}11\cdot11\cdot35 . \supset$

$\vdash :. (\exists . r, y) . \alpha = \iota\text{‘}x . \beta = \iota\text{‘}y . \supset : \alpha \cup \beta \varepsilon 2 . \equiv . \alpha \cap \beta = \Lambda$ (2)

$\vdash . (2) . {*}11\cdot54 . {*}52\cdot1 . \supset \vdash . \text{Prop}$

From this proposition it will follow, when arithmetical addition has been defined, that 1 + 1 = 2.’

Infinity and the Sublime

‘The silence of infinite space terrifies me.’

Pascal

‘Another source of the sublime, is *infinity*. Infinity has a tendency to fill the mind with that sort of delightful horror, which is the most genuine effect, and truest test of the sublime. There are scarce any things which can become the objects of our senses that are really, and in their own nature infinite. But the eye not being able to perceive the bounds of many things, they seem to be infinite, and they produce the same effects as if they really were so. We are deceived in the like manner, if the parts of some large object are so continued to any indefinite number, that the imagination meets no check which may hinder its extending them at pleasure . . .’

Edmund Burke

‘No priestly dogmas, invented on purpose to tame and subdue the rebellious reason of mankind, ever shocked common sense more than the doctrine of the infinite divisibility of extension, with its consequences; as they are pompously displayed by all geometricians and metaphysicians, with a kind of triumph and exultation. A real quantity, infinitely less than any finite quantity, containing quantities infinitely less than itself, and so on *in infinitum*; this is an edifice so bold and prodigious, that it is too weighty for any pretended demonstration to support, because it shocks the clearest and most natural principles of human reason. But what renders the matter more extraordinary, is, that these seemingly absurd opinions are supported by a chain of reasoning, the clearest and most natural; nor is it possible for us to allow the premises without admitting the consequences . . . Reason here seems to be thrown into a kind of amazement and suspence, which, without the suggestions of any sceptic, gives her a diffidence of herself, and of the ground on which she treads.'

David Hume

The mathematician's need to communicate

'The mathematician, the astronomer, the mechanician, sees few men who have much sympathy with his pursuits, or who do not look with indifference on the objects which he pursues. The *world*, to him, consists of a few individuals, by the censures or approbation of whom the public opinion must be finally determined; with them it is material that he should have more frequently intercourse than could be obtained by casual rencounter; and he feels that the society of men engaged in pursuits similar to his own, is a necessary *stimulus* to his exertions. Add to this, that such societies become centers in which information concerning facts is collected from all quarters. For all these reasons, the greatest benefit has resulted from the scientific institutions which, since the middle of the seventeenth century, have become so numerous in Europe.'

John Playfair (1748–1819)

The generosity of algebra

'Algebra is generous, she often gives more than is asked of her.'

D'Alembert

'One cannot escape the feeling that these mathematical formulas have an independent existence and an intelligence of their own, that they are wiser than we are, wiser even than their discoverers, that we get more out of them than was originally put into them.'

Heinrich Hertz, the discoverer of radio waves

Mathematical pictures in physics

Eugene Wigner (page 114) is not the only scientist to have wondered at the amazing capacity of mathematics to represent the physical universe. This is Richard Feynman's meditation on the theme, while accepting his Nobel prize for Physics in 1965, followed by his thoughts on how to go about fitting mathematics to physical reality.

'I would like to interrupt here to make a remark. The fact that electrodynamics can be written in so many ways – the differential equations of Maxwell, various minimum principles with fields, minimum principles without fields, all different kinds of ways, was something I knew, but I have never understood. It always seems odd to me that the fundamental laws of physics, when discovered, can appear in so many different forms that are not apparently identical at first, but, with a little mathematical fiddling you can show the relationship. An example of that is the Schrödinger equation and the Heisenberg formulation of quantum mechanics. I don't know why this is – it remains a mystery, but it was something I learned from experience. There is always another way to say the same thing that doesn't look at all like the way you said it before. I don't know what the reason for this is. I think it is somehow a representation of the simplicity of nature. A thing like the inverse square law is just right to be represented by the solution of Poisson's equation, which, therefore, is a very different way to say the same thing that doesn't look at all like the way you said it before. I don't know what it means, that nature chooses these curious forms, but maybe that is a way of defining simplicity. Perhaps a thing is simple if you can describe it fully in several different ways without immediately knowing that you are describing the same thing.

Many different physical ideas can describe the same physical reality. Thus, classical electrodynamics can be described by a field view, or an action at a distance view, etc. Originally, Maxwell filled space with idler wheels, and Faraday with field lines, but somehow the Maxwell equations themselves are pristine and independent of the elaboration of words attempting a physical description. The only true physical description is that describing the experimental meaning of the quantities in the equation – or better, the way the equations are to be used in describing experimental observations. This being the case perhaps the best way to proceed is to try to guess equations, and disregard physical models or descriptions. For example, McCullough guessed the correct equations for light propagation in a crystal long before his colleagues using elastic models could make head or tail of the phenomena, or again, Dirac obtained his equation for the description of the electron by an almost purely mathematical proposition. A simple physical view by which all the contents of this equation can be seen is still lacking.

Therefore, I think equation guessing might be the best method to proceed to obtain the laws for the part of physics which is presently unknown. Yet, when I was much younger, I tried this equation guessing and I have seen many students try this, but it is very easy to go off in wildly incorrect and impossible directions. I think the problem is not to find the *best* or most efficient method to proceed to a discovery, but to find any method at all. Physical reasoning does help some people to generate suggestions as to how the unknown may be related to the known. Theories of the known, which are described by different physical ideas may be equivalent in all their predictions and are hence scientifically indistinguishable. However, they are not psychologically identical when trying to move from that base into the unknown. For different views suggest different kinds of modifications which might be made and hence are not equivalent in the hypotheses one generates from them in one's attempt to understand what is not yet understood . . .'

The abstract becomes familiar

'Men of science have to deal with extremely abstract and general conceptions. By constant use and familiarity, these, and the relations between them, become just as real and external as the ordinary objects of experience, and the perception of new relations among them is so rapid, the correspondence of the mind to external circumstances so great, that a real scientific sense is developed, by which things are perceived as immediately and truly as I see you now.'

W. K. Clifford (1845–1879)

Gödel becomes an American citizen

'After many years of residence in the United States, the time came for him to take on American citizenship. This required him to answer a number of simple questions about the American Constitution in order to demonstrate his general knowledge and appreciation of it. Moreover, he needed two nominees to vouch for his character and accompany him to this oral examination before a local judge. Gödel's sponsors were impressive – Albert Einstein, who needs no introduction, and Oskar Morgenstern, a famous mathematical economist and co-inventor with John von Neumann of "game theory". Einstein tells the story of how he and Morgenstern became increasingly worried about Gödel's instability and lack of common sense in the run-up to this simple citizenship interview. Apparently, Gödel called Morgenstern on the eve of the interview to tell him that he had discovered a logical loophole in the framing of the Constitution which would enable a dictatorship to be created. Morgenstern told him that this was absurdly unlikely and under no circumstances should he even mention the possibility at his interview the following day. When the day of the interview came, Einstein and Morgenstern tried to distract Gödel from thinking too much about what was in store by generating a steady stream of jokes and stories, hoping that he would be content to turn up, mouth a few rote answers and pleasant platitudes, and depart with his citizenship. John Casti's account of what actually transpired at the interview confirms the distinguished witnesses' worst fears:

At the interview itself the judge was suitably impressed by the sterling character and public personas of Gödel's witnesses, and broke with tradition by inviting them to sit in during the exam. The judge began by saying to Gödel, "Up to now you have held German citizenship." Gödel corrected this slight affront, noting that he was Austrian. Unfazed, the judge continued, "Anyhow, it was under an evil dictatorship . . . but fortunately that's not possible in America." With the magic word dictatorship out of the bag, Gödel was not to be denied, crying out, "On the contrary, I know how that can happen. And I can prove it!" By all accounts it took the efforts of not only Einstein and Morgenstern but also the judge to calm Gödel down and prevent him from going into a detailed and lengthy discourse about his "discovery".'

The Captain's Story

'We have before us an anecdote communicated to us by Captain Basil Hall, a naval officer, distinguished for the extent and variety of his attainments, which shows how expressive such results may become in practice. He sailed from San Blas, on the west coast of Mexico, and after a voyage of eight thousand miles, occupying eighty-nine days, arrived off Rio de Janeiro; having in this interval passed through the Pacific Ocean, rounded Cape Horn, and crossed the South Atlantic, without making any land, or even seeing a single sail, with the exception of an American whaler off Cape Horn. Arrived within a week's sail of Rio, he set seriously about determining, by lunar observations, the precise line of the ship's course, and its situation in it, at a determinate moment; and having ascertained this within from five to ten miles, ran the rest of the way by those more ready and compendious methods, known to navigators, which can be safely employed for short trips between one known point and another, but which cannot be trusted in long voyages, where the moon is the only sure guide.

The rest of the tale, we are enabled, by his kindness, to state in his own words: – "We steered towards Rio de Janeiro for some days after taking the lunars above described, and having arrived within fifteen or twenty miles of the coast, I hove-to at four in the morning, till the day should break, and then bore up; for although it was very hazy, we could see before us a couple of miles or so. About eight o'clock it became so foggy, that I did not like to stand in further, and was just bringing the ship to the wind again, before sending the people to breakfast, when it suddenly cleared off, and I had the satisfaction of seeing the great Sugar-Loaf Rock, which stands on one side of the harbour's mouth, so nearly right ahead that we had not to alter our course above a point in order to hit the entrance of Rio. This was the first land we had seen for three months, after crossing so many seas, and being set backwards and forwards by innumerable currents and foul winds." The effect on all on board might well be conceived to have been electric; and it is needless to remark how essentially the authority of a commanding officer over his crew may be strengthened by the occurrence of such incidents, indicative of a degree of knowledge and consequent power beyond their reach.'

Admirable effects

Henry van Etten explaining in the preface to his *Mathematical Recreations* why he did not give solutions to the problem posed:

'First, that I place not the speculative demonstration with all these problemes, but content my selfe to show them as at the fingers end . . . those which understand the mathematicks can conceive them easily; others for the most part will content themselves onely with the knowledge of them, without seeking the reasons.

Secondly [these things] ought to be concealed as much as they may, in the subtilitie of the way: for that which doth ravish the spirits is, an admirable effect whose cause is unknown: which if it were discovered, halfe the pleasure is lost.'

An Indian theorem?

An Indian chief had three wives, who were preparing to give birth, one on a buffalo hide, the second on a bear hide, and the third on a hippopotamus hide. In due course, the first gave him a son, the second a daughter, and the third, twins, a boy and girl, thereby illustrating the well known theorem that the squaw on the hippopotamus is equal to the sum of the squaws on the other two hides.

John Horton Conway

'Thomas Alva Edison said that genius is one percent inspiration and ninety-nine percent perspiration. If Conway's genius is more than one percent inspiration, then it's because he adds up to more than one hundred percent! He does thousands of calculations, looks at thousands of special cases, until he exposes the hidden pattern and divines the underlying structure. He has drawn (or made out of strings of beads!) tens of thousands of knots, leading to two new ways of looking at their classification and enabling him to push this further than anyone else.

His discovery of the game of Life was effected only after the rejection of many patterns, triangular and hexagonal lattices as well as square ones, and of many other laws of birth and death, including the introduction of two and even three sexes. Acres of squared paper were covered, and he and his admiring entourage of graduate students shuffled poker chips, foreign coins, cowrie shells, Go stones or whatever came to hand, until there was a viable balance between life and death. In the final version a live cell in a rectangular array survives if there are just 2 or 3 live neighbors (a chess king move away) and a dead cell comes to life if it has exactly 3 live neighbors. The implications of this simple set of rules exceed your wildest guesses: Life can simulate a Minsky machine, so Life is universal!

Conway is incredibly untidy. The tables in his room at the Department of Pure Mathematics and Mathematical Statistics in Cambridge are heaped high with papers, books, unanswered letters, notes, models, charts, tables, diagrams, dead cups of coffee and an amazing assortment of bric-à-brac, which has overflowed most of the floor and all of the chairs, so that it is hard to take

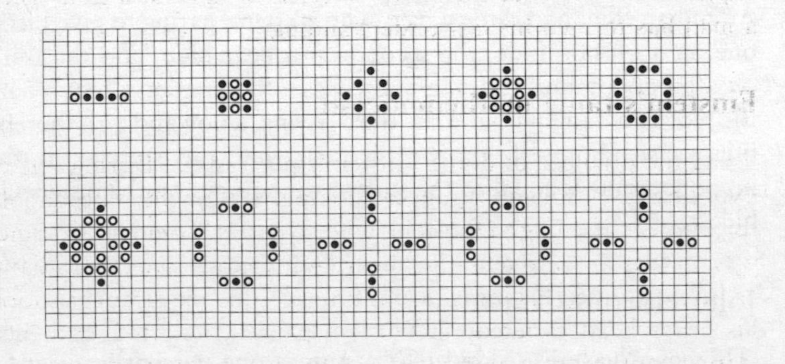

A typical sequence of positions in Conway's Game of Life. A 'line of five' is transformed into the perpetually flashing 'traffic lights' configuration.

more than a pace or two into the room and impossible to sit down. If you can reach the blackboard there is a wide range of colored chalk, but no space to write. His room in college is in a similar state. In spite of his excellent memory he often fails to find the piece of paper with the important result that he discovered some days before, and which is recorded nowhere else. Even Conway came to see that this was not a desirable state of affairs, and he set to work designing and drawing plans for a device which might induce some order amongst the chaos. He was about to take his idea to someone to get it implemented, when he realized that just what he wanted was standing, empty, in the corner of his room. Conway had invented the filing cabinet!'

Newton's confidence

'John Conduitt, a personal friend of Newton, tells the following: Mr Molyneux related to us that after he and Mr Graham and Dr Bradley had put up a perpendicular telescope at Kew, to find out the parallax of the fixed stars, they found a certain nutation of the Earth which they could not account for, and which Molyneux told me he thought destroyed entirely the Newtonian system; and therefore he was under the greatest difficulty how to break it to Sir Isaac. And when he did break it by degrees, in the softest manner, all Sir Isaac said in answer was, when he had told him his opinion, "It may be so, there is no arguing against facts and experiments," so cold was he to all sense of fame at a time when a man has formed his last understanding.'

Einstein's tame mathematician

'I had the privilege of doing Einstein's portrait . . . During one of two sittings a solemn stranger looking, I thought, like an old tortoise, sat listening to Einstein, who, as far as I could understand, was putting forward tentative theories, his expressive face radiant, as he expounded his ideas. From time to time the stranger shook his heavy head, whereupon Einstein paused, reflected, and then started another train of thought. When I was leaving, the presence of a third party was explained. "He is my mathematician," said Einstein, "who examines problems which I put before him, and checks their validity. You see, I am not myself a very good mathematician." '

William Rotherstein

Mathematics and the military

'Mathematics is concerned with the "forms and draughts of all figures, greatness of all bodies, all manner of measures and weights, the cunning working of all tools; with all artificial instruments whatsoever. All engines of war . . . exosters, sambukes, catapults, testudoes, scorpions, petards, grenades, great ordinances of all sorts. By the benefit likewise of geometry, we have our goodly ships, galleys, bridges, mills . . . roofs and arches . . . clocks, curious watches, kitchen jacks, even the wheelbarrow." '

Henry Peacham (1622)

'Certainly, it may be taken as a general rule that the more any person is knowing in mathematics, the fitter he is for all parts of military affairs; and likewise a nation never receives greater damage and disgrace than when ignorant and unfaithful persons are put in places of trust, which I never wish to see in England whilst I am.'

Robert Anderson (1691)

Compare,

'To [David Kinairdy] it was, when he was working as Savilian professor at Oxford, that old Kinairdy confided a model of an improved cannon, which in his enthusiasm to improve the munitions of war, he had designed in his peaceful home by the Deveron. His son, who thought it most ingenious, showed it to Sir Isaac Newton, and the great philosopher evidently agreed with him; but to invent an instrument, the only object of which was to kill better than any cannon in use, seemed to him a fearful abuse of ingenuity. The horrors of Marlborough's wars, where men were slaughtered by the thousand, were they not enough as it was? Who could deserve mercy from his Maker if he were to bid god-speed to such a terrible machine? Sir Isaac asked the professor to destroy the model, which he did, and the little toy which may have been a gatling gun, for aught we know, was broken in pieces.'

Dirac on poetry

'Oppenheimer was working at Göttingen and the great mathematical physicist, Dirac, came to him one day and said: "Oppenheimer, they tell me you are writing poetry. I do not see how a man can work on the frontiers of physics and write poetry at the same time. They are in opposition. In science you want to say something that nobody knew before, in words which everyone can understand. In poetry you are bound to say . . . something that everybody knows already in words that nobody can understand." '

That's showing 'em

'Mrs Maria Drew, a 51-year-old housewife from Waterloo, Iowa, typed out every number from one to one million after her son's teacher told him it was impossible to count up to one million. It took her 5 years and 2,473 sheets of typing paper.'

The Moore method

R. L. Moore (1882–1974) was a Texan topologist and a big man in every way. He was famous for inventing the Moore method of teaching mathematics. Here Paul Halmos expresses his admiration, and his reservations.

'Moore felt the excitement of mathematical discovery and he understood the relation between that and the precision of mathematical expression. He could communicate his feeling and his understanding to his students, but he seemed not to know or care about the beauty, the architecture, and the elegance of mathematics and of mathematical writing. Most of his students inherited his failings as well as his virtues (diluted, of course); only the greatest, such as Wilder and Bing, could overcome the handicap of being a Moore student and become genuine mathematicians.

He was a Texan, almost the fictional prototype. He spoke Texan, he was politically rigid, he had strong prejudices, he stood up when a lady entered the room, and (the story goes) he wouldn't accept students who were black, or female, or foreign, or Jewish. (At least a part of that is false: he had women Ph.D. students, notably Mary Ellen Estill Rudin. So far as I know he did not have a black student.)

Moore, the educated well-spoken Texan mathematician extraordinary, was a hero of mine; Moore, the mathematically outmoded and ethnically prejudiced reactionary power, was a villain.

An effective hero, a productive villain, everyone must admit. He turned out a record-breaking number of Ph.D.'s in mathematics; they loved him and imitated him as far as they could. He did it by what has come to be called the Moore method. It is also called the Texas or Socratic or discovery or do-it-yourself method.

At the first meeting of the class Moore would define the basic terms and either challenge the class to discover the relations among them, or, depending on the subject, the level, and the students, explicitly state a theorem, or two, or three. Class dismissed. Next meeting: "Mr Smith, please prove Theorem 1. Oh, you can't? Very well, Mr Jones, you? No? Mr Robinson? No? Well, let's skip Theorem 1 and come back to it later. How about Theorem 2, Mr Smith?" Someone almost always could do something. If not, class dismissed. It didn't take the class long to discover that Moore really meant it, and presently the students would be proving theorems and watching the proofs of others with the eyes of eagles. One of the rules was that you mustn't let anything wrong get past you – if the one who is presenting a proof makes a mistake, it's your duty (and pleasant privilege?) to call attention to it, to supply a correction if you can, or, at the very least, to demand one.

The procedure quickly led to an ordering of the students by quality. Once that was established, Moore would call on the weakest student first. That had two effects: it stopped the course from turning into an uninterrupted series of lectures by the best student, and it made for a fierce competitive attitude in the class – nobody wanted to stay at the bottom. Moore encouraged competition. Do not read, do not collaborate – think, work by yourself, beat the other guy. Often a student who hadn't yet found the proof of Theorem 11 would leave the room while someone else was presenting the proof of it – each student wanted to be able to give

Moore his private solution, found without any help. Once, the story goes, a student was passing an empty classroom, and, through the open door, happened to catch sight of a figure drawn on a blackboard. The figure gave him the idea for a proof that had eluded him till then. Instead of being happy, the student became upset and angry, and disqualified himself from presenting the proof. That would have been cheating – he had outside help!'

Paul Halmos

Benjamin Banneker

'There is so much to admire in the life of Benjamin Banneker (1731–1806). He was the first American Negro mathematician; he published a very meritorious almanac from 1792 to 1806, making his own astronomical calculations; using a borrowed watch as a model, he constructed entirely from hard wood a clock that served as a reliable timepiece for over twenty years; he won the enthusiastic praise of Thomas Jefferson, who was then the Secretary of State; he served as a surveyor on the Commission appointed to determine the boundaries of the District of Columbia; he was known far and wide for his ability in solving difficult arithmetical problems and mathematical puzzles quickly and accurately. These achievements are all the more remarkable in that he had almost no formal schooling and was therefore largely self-taught, studying his mathematics and astronomy from borrowed books while he worked for a living as a farmer.

But laudable as all the accomplishments of Benjamin Banneker mentioned above are, there is a further item that perhaps draws stronger applause. In his almanac of 1793, he included a proposal for the establishment of the office of Secretary of Peace in the President's Cabinet, and laid out an idealistic pacifist plan to insure national peace. Every country in the world has the equivalent of a Secretary of War. Had Benjamin Banneker's proposal been sufficiently heeded, the United States of America might have been the first country to have a Secretary of Peace! The possibility of realizing this honor still exists – and the time for it is overripe.'

Verse and worse

Multiplication is vexation
Division is as bad
The Rule of Three it puzzles me
And Practice drives me mad.

A possible reason for this confusion is hinted at in the following couplet:

Minus times minus is always plus,
The reasons for this we do not discuss.

More generally:

Seven plagues had Egypt,
We have only three,
Algebra, Arithmetic,
And Plane Geometry.

The language of art

Georges Vantongerloo (1886–1965) was the only sculptor in the De Stijl group. His construction titled '$y = 2x^3 - 13, 5x^2 + 21x$' (1935) consists of three vertical rectangular bars joined by small horizontal rectangular blocks. As a modern critic writes of two of his abstract sculptures, 'Construction $s = r/3$' (1933–4) and 'Group $y = ax^2 + bx + c$' (1931), 'The work titles again speak for themselves,' though not (we will add) in any language familiar to most readers of this book.

Van Etten's Recreations

Henry van Etten was the author of *Mathematical Recreations, Or a Collection of Sundrie excellent Problemes out of ancient and modern Phylosophers, Both useful and Recreative*, published in London, in translation, in 1633. His contents listing illustrates why the word 'mathematics' is plural, not singular:

'Arithmeticke, Geometrie, Cosmographie, Horalographie, Astronomie, Navigation, Musicke, Opticks, Architecture, Staticke, Mechanicks, Chimestrie, Waterworkes, Fireworkes.'

He justified the inclusion of fireworks thus:

'And to them we have also added our Pyrotechnie, knowing that Beasts have for their object only the surface of the earth; but hoping that thy spirit which followeth the motion of fire, will abandon the lower Elements and cause thee to lift up thine eyes to soare in a higher Contemplation, having so glittering a canopie to behould; and these pleasant and recreative fires ascending may cause thy affections also to ascend. The whole whereof we send forth to thee, that desirest the scrutability of things; Nature having furnished us with matter, thy spirit may easily digest them, and put them finely in order, though now in disorder.'

Down with Maths

'MATHS MASTERS

$$\frac{a \times b\,(c-d)}{d \times c\,(b-a)} = \frac{pq + rs}{xg - nbg}$$

The above is what maths masters thrive on and explanes why they are so very stern strict and fearsome. noone in a class ever stirs as a maths master approche you can hear a pin drop and no wonder when you think of the above sum which is enuff to silence anebode.

The only way with a maths master is to hav a very worred xpression. Stare at the book intently with a deep frown as if furious that you canot see the answer. at the same time scratch the head with the end of the pen. After 5 minits it is not safe to do nothing any longer. Brush away all the objects which hav fallen out of the hair and put up hand.

"Sir?" (*whisper*)

"Please sir?" (*louder*)

"Yes, molesworth?" sa maths master. (*Thinks: it is that uter worm agane chiz*)

"Sir i don't quite *see* this."

nb it is esential to sa you don't quite '*see*' sum as this means you are only temporarily bafled by unruly equation and not that you don't kno the fanetest about any of it. [*Dialog continue*:]

"What do you not see molesworth?" sa maths master (*Thinks: a worthy dolt who is making an honest efort*)

"number six sir i can't make it out sir."

"What can you not make out molesworth?"

"number six sir."

"it is all very simple molesworth if you had been paing atention to what i was saing at the beginning of the lesson. Go back to your desk and *think*."

This gets a boy nowhere but it show he is KEEN which is important with maths masters.

Maths masters do not like neck of any kind and canot stand the casual approach.

HOW NOT TO APPROACH A MATHS MASTER

"Sir?"

"Sir sir please?"

"Sir sir please sir?"

"Sir sir please sir sir please?"

"Yes molesworth?"

"I simply haven't the fogiest about number six sir."

"Indeed, molesworth?"

"It's just a jumble of letters sir i mean i kno i couldn't care less whether i get it right or not but what sort of an ass sir can hav written this book."

(*Maths master give below of rage and tear across room with dividers. He hurl me three times round head and then out of the window.*)

Maths masters do not stop at arith and algy they include geom and to do this they hav a huge wooden compass with chalk in the end for the blakboard. The chalk make a friteful noise which set our delicate nerves jangling my dear but this is better than doing the acktual geom itself.

Pythagoras as a mater of fact is at the root of all geom. Insted of growing grapes figs dates and other produce of greece Pythagoras aplied himself to triangles and learned some astounding things about them which hav been inflicted on boys ever since.

Whenever he found a new thing about a triangle Pythagoras who had no shame jumped out of his bath and shouted "Q.E.D."

through the streets of athens its a wonder they never locked him up.'

Geoffrey Willans

Light on his feet

'In 1958, I was visiting Professor at the University of Colorado at Boulder. One day the phone rang, and a woman's voice said, "I am Ann Lee and I want to give you a chance of winning 45 dollars." I said, "Oh." She then asked me, "What was the oldest dance in the world?" I said, "This is a difficult question and I don't know." She then asked me, "Where does the tango come from?" I said, "South America." "Good," she said, "you have answered the question, you are now entitled to 45 dollars of free dancing lessons." You may think for a moment what reply you would make to this, but I would get top marks. I asked her, "Who was the first President of the U.S.A.?" "George Washington," she said. "Good," I replied, "you have won your 45 dollars back again." '

L. J. Mordell

A question of definition

'In California, Bill Honig, the Superintendent of Public Instruction, said he thought the general public should have a voice in defining what an excellent teacher should know. "I would not leave the definition of math," Dr Honig said, "up to the mathematicians." '

The New York Times, 22 October 1985

Gauss on the Higher Arithmetic

'The questions of higher arithmetic often present a remarkable characteristic which seldom appears in more general analysis, and increases the beauty of the former subject. While analytic investigations lead to the discovery of new truths only after the fundamental principles of the subject (which to a certain degree open the way to these truths) have been completely mastered; on the contrary in arithmetic the most elegant theorems frequently arise experimentally as the result of a more or less unexpected stroke of good fortune, while their proofs lie so deeply embedded in the darkness that they elude all attempts and defeat the sharpest inquiries. Further, the connection between arithmetical truths which at first glance seem of widely different nature, is so close that one not infrequently has the good fortune to find a proof (in an entirely unexpected way and by means of quite another inquiry) of a truth which one greatly desired and sought in vain in spite of much effort. These truths are frequently of such a nature that they may be arrived at by many distinct paths and that the first paths to be discovered are not always the shortest. It is therefore a great pleasure after one has fruitlessly pondered over a truth and has later been able to prove it in a round-about way to find at last the simplest and most natural way to its proof.'

The education of Harish-Chandra

'[Harish-Chandra] came to mathematics relatively late and, in spite of enthusiastic initial attempts, there were broad domains of mathematics that he never assimilated in any serious way, although he learned all that he needed ... it is not too much of an exaggeration to say that he manufactured his own tools as the need arose, and that one of the grand mathematical theories of this century has been constructed with the skills with which one leaves a course in advanced calculus.'

He wrote himself:

'I have often pondered over the roles of knowledge or experience, on the one hand, and imagination or intuition, on the other, in the process of discovery. I believe that there is a certain fundamental conflict between the two, and knowledge, by advocating caution, tends to inhibit the flight of imagination. Therefore, a certain naivete, unburdened by conventional wisdom, can sometimes be a positive asset.'

Mathematics from prison

One great mathematical book came out of Napoleon's Russian campaign. Poncelet (1788–1867) became a Russian prisoner of war, and while in captivity he drafted his *Treatise on the Projective*

Properties of Figures. André Weil spent the first few months of the Second World War imprisoned by the Finns as a spy, and in danger of his life. Repatriated to France, he was then thrown into prison in Rouen by the French as a deserter, and again nearly shot. In this letter to his wife he describes his mathematical work in Rouen.

'My mathematics work is proceeding beyond my wildest hopes, and I am even a bit worried – if it's only in prison that I work so well, will I have to arrange to spend two or three months locked up every year? In the meantime, I am contemplating writing a report to the proper authorities, as follows: "To the Director of Scientific Research: Having recently been in a position to discover through personal experience the considerable advantages afforded to pure and disinterested research by a stay in the establishments of the Penitentiary System, I take the liberty of, etc. etc." [. . .]

As for my work, it is going so well that today I am sending Papa Cartan a note for the *Comptes-Rendus*. I have never written, perhaps never even seen, a note in the *Comptes-Rendus* in which so many results are compressed into such a small space. I am very pleased with it, and especially because of where it was written (it must be a first in the history of mathematics) and because it is a fine way of letting all my mathematical friends around the world know that I exist. And I am thrilled by the beauty of my theorems.'

The Romance of Mathematics

'Karl Schellbach, the romantic poet and famous mathematics professor at the Friedrich Wilhelm Gymnasium in Berlin, considered the teaching of mathematics to be a religious vocation, and he believed that mathematicians were priests who should expose as many people as possible to the realms of mathematical blessedness and glory. He felt that both the gifted and the intellectually poor should take part in this kingdom of heaven. The Minister of Culture, von Bethmann-Hollweg, grandfather of the fifth Reichskanzler, once commented that Schellbach's teaching was "an inspired hymn of mathematics". In an effort to expose inexperienced teachers to the mathematical instruction of Schellbach, the ministry introduced a special seminar to be led by the renowned professor. A great number of young mathematicians were introduced to the art of teaching by "old Schellbach" at these seminars. Many of these young men went on to become truly outstanding secondary school teachers . . .

Novalis (1772–1801) preceded Schellbach in proclaiming mathematics to be a religion, and mathematicians to be the blessed priests of the religion. He once commented: "The life of God is mathematics; all divine ambassadors must be mathematicians. Pure mathematics is religion. Mathematicians are the only blessed people." '

Derision by repeated subtraction

'There used to be an admissions examination, the "11-plus", to British secondary grammar schools. A question asked on one occasion was, "Take 7 from 93 as many times as you can." One child answered, "I get 86 every time!" '

Boy from the word go

'*I am interested in the fact that you are the second of three daughters of J. L. Synge. I have read that women scientists are quite frequently second daughters. The explanation given is that a second daughter feels that she should have been a boy, so there's a tendency for her to take after the father and try to become a son in her interests. I don't know whether that is true.*

MORAWETZ: I was the boy in the family. I was the boy from the word go. But that doesn't mean that I had an especially close relationship with my father. My older sister really had a closer relationship. For instance, when we were girls and we wanted to do something he didn't want us to do, she was always delegated to negotiate with him. I would say rather that I was in competition with my father. Better not print that! [*Laughs.*] That was the pattern, quite different. But I was definitely the boy. In fact, it was rather funny. When I was seven and we returned to Canada from Ireland, I had had a major operation on my leg so I had my leg in a brace. It had been arranged that my father was going to look after me, and my mother was going to look after my little sister, who was just a baby. My older sister was going to look after herself. It was all very formally arranged. And the first thing my father did was to take me to have my hair cut like a boy's. This was supposedly because it was a mess and he didn't want to have to take care of it, but that's what he did. Of course that was 1930, and that was a time when people did have their hair cropped. But when the people on the ship said I was a little boy, I was very unhappy.

Did your mother have any mathematical interest?

MORAWETZ: She went to Trinity College and studied mathematics, but her brother persuaded her that there was no future in math for her, so she switched to history. At one point, when she was ready for secondary school, there was a competition among the Protestant girls in Ireland and she was placed first. Actually she won a prize to go to the conservatory of music in London, but her mother wouldn't let her go. She went to Trinity College instead, and that's where she and my father met. After they were married – he was still in school then – she taught to support them, so she never got a degree.'

Cathleen S. Morawetz (*b.* 1923)

Kicks

Here is the puzzle composer, Hubert Phillips, giving his criterion for a good puzzle:

'Does its statement involve, not only the labour of working out the answer (which for many has a very slight appeal) but also the excitement of first discovering how the answer is to be arrived at? My main pleasure, in constructing puzzles, lies in seeking to provide just this "kick".'

Compare this quotation from the leading English mathematician of the period, G. H. Hardy:

'. . . the puzzle columns of the popular newspapers. Nearly all their immense popularity is a tribute to the drawing power of rudimentary mathematics . . . They know their business; what the public wants is a little intellectual "kick", and nothing else has quite the kick of mathematics . . . The fact is that there are few more "popular" subjects than mathematics. Most people have some appreciation of mathematics, just as most people can enjoy a pleasant tune.'

Euler becomes blind

'In his last years in Petersburg Euler had more free time for mathematics than ever before. He very soon lost the sight of his one remaining eye. Like Bach, he underwent the torment of an operation for cataract, which was unsuccessful and rendered him almost totally blind. If anything, this enforced end to most of the ordinary duties of life left him still freer to work. About half of his 800 publications were written in these, the last seventeen years of his life. In 1766, the year he moved, Euler composed the first general treatise on hydrodynamics; it was to be about one hundred years before anyone wrote another. The next year Euler wrote his famous *Complete Introduction to Algebra*. After Euclid's *Elements*, this is the most widely read of all books on mathematics, having been printed at least thirty times in three editions and in six languages; selections were being used as textbooks in the Boston schools seventy years afterward. The next year, 1768, Euler wrote his three-volume treatise on geometrical optics and his tract on the motion of the moon; both of these are filled with colossal calculations, and the latter contains a single table 144 pages long. In 1770 he wrote a monograph on the difficult orbit of a comet which had appeared the year before.

Euler's total blindness put an end to composition of such long treatises, and the great increase in the annual number of his publications reflects the change in his method of work. In the middle of his study he had a large table with a slate top. Being barely able to distinguish white from black, he could write a few large equations. Every morning a young Swiss assistant read him the post, the newspaper, and some mathematical literature. Euler then explained some problem he had been sleeping on and proposed a method of attacking it. The assistant was usually able to produce the outline for a draught of a short memoir, or part of one, by the next morning. In 1775, for example, Euler composed more than one complete paper per week; these run from ten to fifty pages in length and concern widely different special problems.

Euler's memory, always remarkable, had by now become

phenomenal. He could still recite the *Aeneid* in Latin from beginning to end, remembering also which lines were first and last on each page of the edition from which he had learned it some sixty years earlier. Enormous equations and vast tables of numbers were ready on demand for the eye of his mind. He became one of the sights of the town for distinguished visitors, with whom he usually spoke on non-mathematical topics. Amazed by the breadth and immediacy of his knowledge concerning every subject of discourse, they spread unbelievable fairy tales about what he could do in his last years.'

A prodigious childhood

Norbert Wiener (1894–1964) graduated from High School at the age of eleven. On the appointed evening, two girls were chosen to stand on either side of him and poke him if he started to fall asleep, because it was past his normal bedtime.

However, he was not a natural prodigy, but the results of his father's deliberate attempts to create a family of brilliant children, or, perhaps, the combination of great natural talent and a bizarre upbringing.

A journalist reported,

'Professor Leo Wiener, of Harvard University . . . believes that the secret of precocious mental development lies in early training . . . He is the father of four children, ranging in age from four to sixteen; and he has had the courage of his convictions in making them the subjects of an educational experiment. The results have . . . been astounding, more especially in the case of his oldest son, Norbert.

This lad, at eleven, entered Tufts College, from which he graduated in 1909, when only fourteen years old. He then entered the Harvard Graduate School . . .'

His father explained:

'Just what method have I used? . . . Instead of leaving them [children] to their own devices – or, worse still, repressing them, as is generally done – they should be encouraged to use their minds to think for themselves, to come as close as they can to the intellectual level of their parents . . .

What is no less important, every child should be carefully studied to determine aptitudes. One child will have a natural bent for mathematics, another for reading, another for drawing, and so forth . . .

Take the case of my boy Norbert. When he was 18 months old, his nurse-girl one day amused herself by making letters in the sand of the seashore. She noticed that he was watching her attentively, and in fun she began to teach him the alphabet. Two days afterward she told me, in great surprise, that he knew it perfectly. Thinking that this was an indication that it would not be hard to interest him in reading, I started teaching him how to spell at the age of three. In a very few weeks he was reading quite fluently, and by six was acquainted with a number of excellent books, including works by Darwin, Ribot, and other scientists, which I had put in his hands in order to instill in him something of the scientific spirit. I did not expect him to understand everything he read.'

Leo Wiener's standards were always very high. When Norbert made mistakes in his studies, his father was often extremely critical and harsh, as he remembered later:

'Algebra was never hard for me, although my father's way of teaching it was scarcely conducive to peace of mind. Every mistake had to be corrected as it was made. He would begin the discussion in an easy, conversational tone. This lasted exactly until I made the first mathematical mistake. Then the gentle and loving father was replaced by the avenger of blood. The first warning he gave me of my unconscious delinquency was a very sharp and aspirated "What!" and if I did not follow this by coming to heel at once, he would admonish me, "Now do this again!" By this time I was weeping and terrified. Almost inevitably I persisted in sin, or what was worse, corrected an admissible statement into a blunder. Then the last shreds of my father's temper were torn . . .

The very tone of my father's voice was calculated to bring me to a high pitch of emotion, and when this was combined with irony and sarcasm, it became a knout with many lashes. My lessons often ended in a family scene. Father was raging, I was weeping and my mother did her best to defend me, although hers was a losing battle. She suggested at times that the noise was disturbing the neighbors and that they had come to the door to complain, and this may have put a measure of restraint on my father without comforting me in the least. There were times for many years when I was afraid that the unity of the family might not be able to stand these stresses, and it is just in this unity that all of a child's security lies.'

Games and situations

'Men are never more ingenious than when they are inventing games; the mind finds satisfaction in this activity . . . After games which depend solely on numbers come games of situation, [after these] come games involving movement . . . Eventually we may hope that there will be a series of games, treated mathematically.'

Leibniz (1646–1716)

Leibniz was fascinated by *situation*, and would have appreciated the eventual development of vector algebra and topology.

'I am still not satisfied with algebra, because it does not give the shortest methods or the most beautiful constructions in geometry. This is why I believe that, so far as geometry is concerned, we need still another analysis which is distinctly geometrical or linear and which will express *situation* [*situs*] directly as algebra expresses *magnitude* directly. And I believe that I have found the way and that we can represent figures and even machines and movements by characters, as algebra represents numbers or magnitudes. I am sending you an essay which seems to me to be important.'

Travelling salesman ants

'To find the shortest round trip to 50 chosen cities in Europe a mathematician would usually recruit a massive computer, a complex program and set aside plenty of time. Researchers at BT, however, found the solution in record time, with a workstation and a collection of "software ants" – autonomous programs a few hundred lines long which, together, can solve enormously difficult problems by dealing with their own simple ones.

BT, which has developed the programs in the past year, says its method could be applied to many problems where a complex series of decisions is needed to achieve the best use of resources. Examples include searching for information on a number of databases, designing circuits on microchips, advising fighter pilots under multiple attack, or sending out telephone engineers to fix faults.

The ants will also help to make software "agents" designed to explore the information superhighways. Peter Cochrane, head of core technologies research at BT, says that with the amount of information that can be accessed from the superhighway doubling every three years the system will have immense value because it will allow people to find information more quickly.

The task of finding the shortest distance between points is known as the travelling salesman problem. The number of possible solutions increases factorially: for four points there are 24 feasible routes, but for 30 there are 2.65×10^{32}.

The previous world best for a proven solution was 3,000 points, which took a Cray supercomputer 18 months to work out. BT's method has solved a problem involving 30,000 points on a workstation in 44 hours, and is accurate to within 4 per cent.

BT's solution to the 50-city problem is like a massive computer game played out on a computer representation of a map of Europe. Each of the ants is assigned a position on the map, and from then on has to obey simple rules. Ants cannot, for example, stray more than a certain distance from their nearest neighbours: they act as if they are all tied together on a long loop of string. If an ant finds a city it may breed, making a copy of itself. And the closer it is to a city, the stronger the attraction to it. If, however, an ant spends too long away from a city, it dies. The computer then generates a new one at some other point between the deceased's nearest neighbours.

To start solving the problem, the computer generates 50 ants joined in a loop and sets them loose to search for cities according to their simple instructions. Eventually they find cities, for which they are rewarded and survive, or they die. Evolution does the rest.

The scientists who devised this method, Shara Amin and José-Luis Fernandez of BT's systems research division at Martlesham Heath near Ipswich, believe they can reach a solution 10 times faster still by making the ants more self-sufficient. At present, the governing program has to inform the ants when they are near a city. Amin and Fernandez want the ants to be able to judge their own positions.'

Paul Valéry and mathematics

Paul Valéry (1871–1945), the French poet and philosopher, had extraordinarily wide interests, including mathematics and physics – his friends included the scientists Paul Langevin and Louis de Broglie.

These passages are by Pierre Féline, who introduced Valéry to mathematics. Valéry was a keen student, though, as he put it, 'I studied mathematics, but in a very odd spirit, as a *model* of acts of the mind,' as one aspect of his life-long interest in the nature of consciousness, mathematics being the type of pure mental activity.

'Fairly often we talked about mathematics; he was still attracted by the theory of groups, and also of groups of transformations. His mind luxuriated in moving in this realm of speculation.

I often teased him about his difficulty in understanding an abstract proposition, or even in stating it without resorting to a concrete example. I made use of a fairly severe test for measuring his capacity for abstraction. It so happened that we both owned a work of uncommon transcendence: Georg Cantor's *Théorie des ensembles*. No vast amount of knowledge was necessary to get through this work; the elementary rules of the calculus sufficed. But alas! After reading a few pages the mind stops, exhausted. The next day we would try to go on, but each day we would advance less and less . . . "What page are you on?" I would ask my friend from time to time. He replied, "I have to reread such and such pages before I can continue . . ." As for myself, where am I today? To what stage of that transcendent construction must I decide to find the strength to advance anew?

His mind was most drawn toward the speculations of the calculus. To create symbols without apparent connection with the real, and to give them life by defining their modes of variation and combination; with these symbols to form groups and consider each form of transformation as a new abstract entity, and to set to studying the groups they can constitute by their structure ... Paul showed keen curiosity for all these degrees of transcendence. When I saw him again later, I found him quite at ease in this domain, where thought sees only transformations and combinations of abstract concepts.'

Grace Chisholm Young (1868–1944)

'Grace Young was a "super woman" before the term was invented. In 1889, having initially studied to be a pianist, she entered Girton, the first college in England dedicated to a university education for women, and began to study mathematics. Five years later, under the direction of Felix Klein in Göttingen, she became the first woman to pass the normal examination in Prussia for the doctor's degree. The following year she married her Girton mathematics tutor, W. H. Young, and inspired him – at the age of forty – to obtain a Göttingen degree as well.

The Youngs settled in Switzerland, where she raised their six children and pursued interests in music, languages, biology, education and writing as well as mathematics. She also completed the course (although she did not take the final examination) for a medical degree. Her husband "commuted" to support the family, fulfilling professional obligations at universities as farflung as Aberystwith and Calcutta, all the while sending home mathematical ideas for his wife to study and develop.

"I hope that you enjoy this working for me," he wrote. "... I feel partly as if I were teaching you, and setting you problems which I could not quite do myself but could enable you to ... The fact is that our papers ought to be published under our joint names, but if this were done neither of us would get the benefit ...'

Their mathematical collaboration (220 papers and articles and several books) was indeed so close that ... historians of the science have treated it as a joint bibliography. Among his contributions to mathematics are his discovery (independent of Lebesgue) of the Lebesgue integral and his work on Fourier series and cluster sets.

Her most important independent work is a group of papers in which she studied derivatives of real functions.

During the Second World War, after the fall of France, the Youngs were inadvertently separated. They both died before the end of the war, and so never saw each other again.'

'A paradox, a paradox, a most ingenious paradox!'

'The theory of probabilities is at bottom only common sense reduced to calculation.'

Pierre Simon Laplace

Laplace was a wonderful mathematician and one of the founders of probability theory, but his confidence in common sense was misplaced. Few subjects contain more counter-intuitive results than probability. Here are three 'paradoxes' (all of which can, of course, be explained).

A medical researcher does a carefully controlled experiment whose result is that new medicine X is more effective on male patients than a placebo. The experiment is then repeated on female subjects, with the same result, medicine X is more effective than a placebo. The data from the two experiments are then added together, and they prove that *overall* medicine X is *less* effective than the placebo. Is this possible? Yes.

We toss a fair coin repeatedly until we score two heads (HH) or a head and a tail (HT) in succession. Obviously the probability that (HH) will occur sooner than (HT) is equal to the probability that (HT) will occur sooner than (HH), since after the initial (H) has been tossed, a following (H) or a following (T) are equally likely.

Yet, in fact, (HT) is more likely to turn up first than (HH). The former occurs on average once in four throws, the latter once every six throws.

Two teams of equal ability are playing against each other. 'Equal ability' means that both teams are equally likely to be the next to score. There is a fifty per cent chance that one team will be in the lead throughout the second half, no matter how many points are scored.

The Unreasonable Effectiveness of Mathematics in the Natural Sciences

Many thinkers have pondered the extraordinary manner in which mathematics 'fits' nature. Why should the world be composed, as it seems to be, on mathematical principles? From the Greeks to the early Renaissance, it was believed in the West that the Heavens were constructed of circles and spheres, these being the most perfect of all forms. When Kepler replaced the circles by ellipses, the puzzle became more difficult to answer, rather than less, and the amazing, even bizarre, success of quantum theory and other branches of modern science seems even more baffling.

Einstein thought that the most incomprehensible thing about the universe is that we can comprehend it. Eugene Wigner, a Nobel prize-winning physicist, found the effectiveness of mathematics in physics and the other sciences nothing short of miraculous. He also wrote the most famous account of the phenomenon, from which this passage is taken.

'It is difficult to avoid the impression that a miracle confronts us here, quite comparable in its striking nature to the miracle that the human mind can string a thousand arguments together without getting itself into contradictions or to the two miracles of the existence of laws of nature and of the human mind's capacity to divine them. The observation which comes closest to an explanation for the mathematical concepts' cropping up in physics which I know is Einstein's statement that the only physical theories which we are willing to accept are the beautiful ones. It stands to argue that the concepts of mathematics, which invite the exercise of so much wit, have the quality of beauty. However, Einstein's observation can at best explain properties of theories which we are willing to believe and has no reference to the intrinsic accuracy of the theory.

The miracle of the appropriateness of the language of mathematics for the formulation of the laws of physics is a wonderful gift which we neither understand nor deserve. We should be grateful for it and hope that it will remain valid in future research and that it will extend, for better or for worse, to our pleasure even though perhaps also to our bafflement, to wide branches of learning . . .'

'Mastermind' in Africa

'Kpelle children play a game based on arrangements of objects. Sixteen stones are arranged in two rows of eight each. One person is sent away, and the others choose a stone. When the "out" person returns, he must determine which stone has been selected. He may ask four times in which of the two rows the stone is located. After each reply, he may rearrange the stones within the two rows. He must be able to identify the chosen stone after the fourth reply.

The key to the solution lies in the procedure by which the stones are rearranged each time. After the first reply the questioner rearranges eight stones, half the original number, so that they are not in different rows; after the second reply he interchanges four stones, half the previous number, and the next time he changes the position of two stones. The answer to the last question determines precisely which stone has been chosen.'

A military approach

'What then are the rules for good mathematical writing? Answer: local clarity and global organisation. In the terminology of another subject: meticulous tactics and sound strategy.'

Paul Halmos

The pain of noncommunication

'. . . it would be astonishing if the reader could identify more than two of the following names: Gauss, Cauchy, Euler, Hilbert, Riemann. It would be equally astonishing if he should be unfamiliar with the names of Mann, Stravinsky, de Kooning, Pasteur, John Dewey. The point is not that the first five are the mathematical equivalents of the second five. They are not. They are the mathematical equivalents of Tolstoy, Beethoven, Rembrandt, Darwin, Freud. The geometry of relativity – the work of Riemann – has had consequences as profound as psychoanalysis. The mathematical equivalents of Mann, Stravinsky, and the others would be Bochner, Thom, Serre, Cartan, and Weil, and it is all but certain that the reader is unfamiliar with every one of these names. There is really no reason he should be. But that is precisely the cruelty of the situation. All professions reward accomplishment in part with admiration by peers, but mathematics can reward it with admiration of no other kind. It is, in fact, impossible for a mathematician even to talk intelligently to non-mathematicians about his mathematical work. In the company of friends, writers can discuss their books, economists the state of the economy, lawyers their latest cases, and businessmen their recent acquisitions, but mathematicians cannot discuss their mathematics at all. And the more profound their work, the less understandable it is; a spirited high-school teacher can regale his audience with puzzles and magic squares, but there is no way for the serious mathematician to talk to the non-mathematician about his latest results on the homotopy groups of spheres.'

Alfred Adler

Gauss's brain

Paul Broca (1824–1880) was a medical professor who founded the Anthropological Society of Paris in 1859. One of early anthropologists' favourite occupations was craniometry, measuring the size of human skulls, and drawing inferences therefrom. Broca firmly believed that human intelligence was related to the physical size of the brain, and that – wait for it! – men were more intelligent than women because they had larger brains, and similarly that more eminent men had larger brains than the less eminent.

'The small brains were troublesome, but Broca, undaunted, managed to account for all of them. Their possessors either died very old, were very short and slightly built, or had suffered poor preservation. Broca's reaction to a study by his German colleague Rudolf Wagner was typical. Wagner had obtained a real prize in 1855, the brain of the great mathematician Karl Friedrich Gauss. It weighed a modestly overaverage 1,492 grams, but was more richly convoluted than any brain previously dissected . . . Encouraged, Wagner went on to weigh the brains of all dead and willing

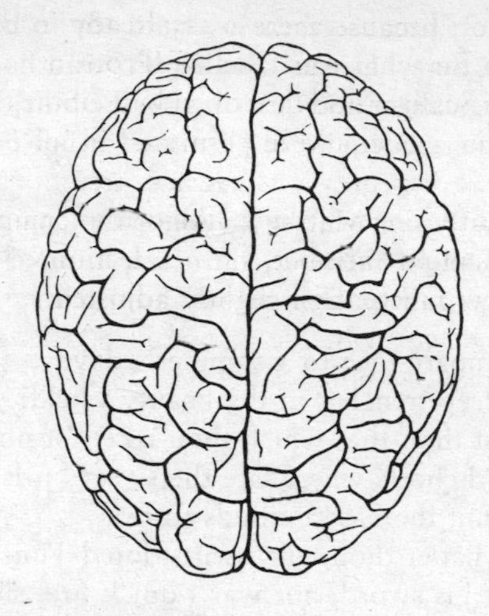

professors at Göttingen [where Gauss's brain is preserved in the anatomical collection], in an attempt to plot the distribution of brain size among men of eminence.'

The Oulipo

The Oulipo, or *Ouvroir de Littérature Potentielle*, was founded in Paris in 1960 by the mathematician François Le Lionnais, author and editor of *Great Currents of Mathematical Thought*, who had been a member of Dada groups in the early 1920s, and the writer Raymond Queneau, who was once a surrealist. It is devoted to experimental literature, more often than not with a mathematical flavour. The combinatorialist Claude Berger and Paul Braffort, a computer scientist and musician, are members; so were Marcel Duchamp, and the novelist Italo Calvino.

When Queneau and Le Lionnais founded the group, Queneau was in the process of composing *Cent mille milliards de poèmes*, which consisted of a set of ten sonnets, which were intended to be read by choosing one of the ten first lines, followed by one of the ten second lines, and so on, giving, since a sonnet has fourteen lines, 10^{14} possible poems.

It might seem that this multiplies by a large factor the total number of poems ever composed in the history of the world. It possibly does not, because there was already in the seventeenth century a mania for what was labelled 'Protean Poetry', invented by Julius Caesar Scaliger and described by Leibniz, in which selections of words in a poem were permuted in all possible combinations.

This is a variant, using Matthew's Algorithm, named after Harry Matthews, an Oulipo member, on a selection of Shakespeare's sonnets, with the punctuation slightly adjusted.

> Shall I compare thee to a summer's day
> And dig deep trenches in thy beauty's field?
> Why lov'st thou that which thou receiv'st not gladly,
> Bare ruin'd choirs where late the sweet birds sang?
> Anon permit the basest clouds to ride
> And do what'er thou wilt, swift-footed Time:
> Nor Mars his sword, nor war's quick fire, shall burn
> Even such a beauty as you master now.
> Love's not Time's fool, though rosy lips and cheeks
> (When other petty griefs have done their spite,
> And heavily) from woe to woe tell o'er
> That Time will come and take my love away;
> For thy sweet love remembered such wealth brings
> As any she belied beyond compare.

A Primer in Humility

It is well known that the Europeans obtained their Arabic numerals from the Arabs who got them from India, and that the Chinese used negative numbers long before anyone in Europe, and that Pascal's Triangle was used by the Chinese long before Pascal studied it, but such facts tend to be dismissed as 'elementary', with the implication that when it comes to more difficult mathematics, Europe was on its own. Not so.

The values of π found by Liu Hui (*c.* AD 200) and Tsu Chung Chih (*c.* AD 400) remained the most accurate known for a thousand years.

The formula for the volume of a truncated triangular prism which is credited in the West to Legendre's *Éléments de géometrie* (1794) appears in the *Chiu Chang Suan Shu*, written between 300 BC and AD 200.

A cubic interpolation formula popularized by Kuo Shou Ching (*c.* 1275) was later known in Europe as the Newton–Stirling formula. The Gregory–Newton interpolation formula was first published by Chu Shih-chieh in his *Precious Mirror of the Four Elements* (1303).

Horner's method for the numerical solution of equations is identical to the Chinese method used over five hundred years earlier. (See page 142.)

Brounker's method (1657) for solving indeterminate equations of the form $ax^2 + bx + c = y$ was discovered by Bhaskaracharya (born 1114).

Euler's *theorem elegantissimum* for solving Pell's equation was used as early as AD 600 by Brahmagupta, who also gave the formulae for the diagonals of a cyclic quadrilateral, first found in Europe by Snell (1619). A fifteenth-century commentary on Bhaskaracharya gives the formula for the radius of a cyclic quadrilateral in terms of its sides, published in Europe by L'Huillier in 1782.

Madhava (*c.* 1340–1425) finds π correct to 11 decimal places. Vieta in 1579 only found 9. Madhava is also credited with what we call the Gregory series for $\tan^{-1} x$, found in 1667. If Madhava was not the discoverer – he may have been given credit by admiring followers – it was still discovered first in Kerala in south India, where it was attributed to Madhava in several texts of the fifteenth century. Madhava, or other Keralese mathematicians, also discovered Leibniz's series for π, and the power series for sine x and cosine x, usually attributed to Newton.

A footnote: Madhava's work is reminiscent of that of another Indian genius, Ramanujan. Where did Ramanujan grow up? At Kumbakonam, not far from Madhava's birthplace.

Cut it out

'The physicist Freeman Dyson relates a conversation between the astronomer James Jeans and the topologist Oswald Veblen in the early 1900s about the Princeton curriculum. "We may as well cut out group theory," said Jeans. "That is a subject that will never be of any use in physics." '

Group theory is central to quantum mechanics.

Gauss and Poincaré: one step ahead of history

Gauss (1777–1855) published his most famous masterpiece, on number theory, the *Disquisitiones Arithmeticae*, in 1801 when he was just 24 years old.

'Perhaps one of the most remarkable parts of the *Disquisitiones* is the section where Gauss defines the composition of two binary quadratic forms and (without knowing what a group is) proves that the classes of binary quadratic forms with given discriminant form a finite group with composition as a group law. In fact he even proves that this group is a direct product of cyclic groups, and . . . Gauss's proof can be generalised to prove that any finite Abelian group is a direct product of cyclic groups!'

Poincaré managed the same feat:

'Poincaré was a precursor of set theory, in the sense that he applied it even before it was born, in one of his most striking and justly celebrated investigations. He showed indeed that the singularities of the automorphic functions form either a whole circle or a Cantor set. This last category was of a kind which his predecessor's imagination could not even conceive. The set in question is one of the most important achievements of set theory, but . . . Cantor himself did not discover it until later.'

Cardan's madness

'. . . we must admit that some parts of the mental process develop so deeply in the unconscious that some parts of it, even important ones, remain hidden from our conscious self. We come very near the phenomena of dual personality such as were observed by psychologists of the nineteenth century.

Even intermediaries seem to have existed between the two kinds of phenomena. I think of Socrates' ideas being suggested to him by a familiar demon or also of the nymph Egeria whom Numa Pompilius used to consult frequently.

An analogous example can possibly be spoken of in the mathematical field. It is Cardan, who is not only the inventor of a well-known joint which is an essential part of automobiles, but who has also fundamentally transformed mathematical science by the invention of imaginaries. Let us recall what an imaginary quantity is. The rules of algebra show that the square of any number, whether positive or negative, is a positive number: therefore, to speak of the square root of a negative number is mere absurdity. Now, Cardan deliberately commits that absurdity and begins to calculate on such "imaginary" quantities.

One would describe this as pure madness; and yet the whole development of algebra and analysis would have been impossible without that fundament – which, of course, was, in the nineteenth century, established on solid and rigorous bases. It has been written that the shortest and best way between two truths of the real domain often passes through the imaginary one.

We have mentioned Cardan's case with Socrates' and Numa Pompilius', because he too is reported by some of his biographers to have received suggestions from a mysterious voice at certain periods of his life. However, testimonies on that point do not agree at least in details.'

'Cardan, called by his contemporaries the greatest of men and the most foolish of children – Cardan, who first dared to criticise Galen, to exclude fire from the number of the elements, and to call witches and saints insane – this great Cardan was the son, cousin, and father of lunatics, and himself a lunatic all his life. "A stammerer, impotent, with little memory or knowledge," he himself wrote, "I have suffered since childhood from hypnofantastic hallucinations." Sometimes it was a cock which spoke to him in a human voice; sometimes Tartarus, full of bones, which displayed itself before him. Whatever he imagined, he could see before him as a real object. From the age of nineteen to that of twenty-six, a genius, similar to one which already protected his father, gave him advice and revealed the future. When he had reached the age of twenty-six he was not altogether deprived of supernatural aid; a recipe which was not quite right forgot one day the laws of gravity, and rose to his table to warn him of the error he was about to commit.

He was hypochondriacal, and imagined he had contracted all the diseases that he read of: palpitation, sitophobia, diarrhoea, enuresis, podagra, hernia – all these diseases vanished without treatment, or with a prayer to the Virgin. Sometimes his flesh smelled of sulphur, of extinguished wax; sometimes he saw flames and phantoms appear in the midst of violent earthquakes, while his friends perceived nothing. Persecuted by every government, surrounded by a forest of enemies, whom he knew neither by name nor by sight, but who, as he believed, in order to afflict and dishonour him, had condemned his much-loved son, he ended by believing himself poisoned by the professors of the University of Pavia, who had invited him for this purpose.

His sensibility was so perverted, that he never felt comfortable except under the stimulus of some physical pain; and in the absence of natural pain, he procured it by artificial means, biting his lips or arms until he fetched blood. "I sought causes of pain to enjoy the pleasure of the cessation of pain, and because I perceived that when I did not suffer I fell into so grave and troublesome a condition, that it was worse than any pain." This fact helps us to understand many strange tortures which madmen have voluptuously imposed on themselves. He had so blind a faith in the revelations of dreams, that he printed a strange work *De Somniis*, conducted his medical consultations, concluded his marriage, and began his works (for example, that on the *Varietà delle Cose* and *Sulle Febbri*) in accordance with dreams.'

Really?

'Thus mathematics may be defined as the subject in which we never know what we are talking about, nor whether what we are saying is true.'

Bertrand Russell

This saying is as famous as it is false: if it were true, then chess players, also, would suffer the same handicap, which they of course do not.

A unique diagram

'The two German mathematicians whom Poincaré compares are Weierstrass and Riemann. That, as he concludes, Riemann is typically intuitive and Weierstrass typically logical is beyond contestation. But as to the latter, Poincaré says, "You may leaf through all his books without finding a figure." It strikes me that there happens to be there an error of fact. It is true that *almost* no memoir of Weierstrass implied any figure: there is only one exception; but there is one, and this exception occurs in one of his most masterly and clear-cut works, one giving the most complete impression of perfection: I mean his fundamental method in the calculus of variations. Weierstrass draws a simple diagram and, after that initial step is taken, everything goes on in the profoundly logical way which is undoubtedly his characteristic, so that, by merely looking at that diagram, anyone sufficiently acquainted with mathematical methods could have rebuilt the whole argument. But of course there was an initial intuition, that of constructing the diagram.'

Hadamard

Minkowski and Smith

'Minkowski [1864–1909] was a chubby-faced boy with a scholar's pince-nez perched rather incongruously on his still unformed nose. While he was in Berlin, he had won a monetary prize for his mathematical work and then given it up in favor of a needy classmate. But this was not known in Königsberg [his home town]. (Even his family learned of the incident only much later when the brother of the classmate told them about it.) Although Minkowski was still only 17 years old, he was involved in a deep work with which he hoped to win the Grand Prix des Sciences Mathématiques of the Paris Academy.

The Academy had proposed the problem of the representation of a number as the sum of five squares. Minkowski's investigations, however, had led him far beyond the stated problem. By the time the deadline of June 1, 1882, arrived, he still had not had his work translated into French as the rules of the competition required. Nevertheless, he decided to submit it. At the last minute, at the suggestion of his oldest brother Max, he wrote a short prefatory note in which he explained that his neglect had been due to the attractions of his subject and expressed the hope that the Academy would not think "I would have given more if I had given less." As an epigraph he inscribed a line from Montaigne: "Rien n'est beau que le vrai, le vrai seul est aimable."

. . . in the spring of 1883, came the announcement that this boy, still only 18 years old, had been awarded jointly with the well-known English mathematician Henry Smith the Grand Prix des Sciences Mathématiques. The impression which the news made in Königsberg can be gauged by the fact that Judge Hilbert admonished David [his son] that presuming an acquaintance with "such a famous man" would be "impertinence".

For a while it seemed, though, that Minkowski might not actually receive his prize. The French newspapers pointed out that the rules of the competition had specifically stated that all entries must be in French. The English mathematicians let it be known that they considered it a reflection upon their distinguished countryman, who had since died, that he should be made to share a mathematical prize with a boy. ("It is curious to contemplate at a distance," an English mathematician remarked some forty years later, "the storm of indignation which convulsed the mathematical circles of England when Smith, bracketed after his death with the then unknown German mathematician, received a greater honor than any that had been paid to him in life.") In spite of the pressures upon them, the members of the prize committee never faltered. From Paris, Camille Jordan wrote to Minkowski: "Work, I pray you, to become a great mathematician." '

Henry John Stephen Smith (1826–1883) was a distinguished Irish-born English mathematician who had published, in 1858, the solution to the very prize problem proposed by the Paris Academy, and which Minkowski had been working on. When Smith read the announcement of the prize, he wrote to Hermite, pointing out his prior publication. Hermite admitted that the Prize Committee had been unaware of his work and suggested that he prepare and submit his memoir in the form demanded by the rules. He did so, his entry was judged, with Minkowski's, to be worthy of the prize, but by then he had died, in February 1883.

An unhappy competition

Felix Klein (1849–1925) was in his earlier years the leading German mathematician of his day. In 1883 he learnt that the brilliant young Frenchman, Henri Poincaré (1854–1912) was attempting to solve the very problems on automorphic functions that he, Klein, was working on. Klein tried desperately to keep up with Poincaré and did so, just, but the result was a mental breakdown which left him in a deep depression. He never recovered his former powers.

Speaking the lingo

'Just as everybody must strive to learn language and writing before he can use them freely for the expression of his thoughts, here too there is only one way to escape the weight of formulae. It is to acquire such power over the tool . . . that, unhampered by formal technique, one can turn to the true problems . . .'

Hermann Weyl

Dirac and women

'Being a great theoretical physicist, Dirac liked to theorize about all the problems of daily life, rather than to find solutions by direct experiment. Once, at a party in Copenhagen, he proposed a theory according to which there must be a certain distance at which a woman's face looks its best. He argued that at $d = \infty$ one cannot see anything anyway, while at $d = 0$ the oval of the face is deformed because of the small aperture of the human eye, and many other imperfections (such as small wrinkles) become exaggerated. Thus there is a certain optimum distance at which the face looks its best.

"Tell me, Paul," I asked, "how close have you ever seen a woman's face?"

"Oh," replied Dirac, holding his palms about two feet apart, "about that close."

Several years later Dirac married "Wigner's sister", so known among the physicists because she was the sister of the noted Hungarian theoretical physicist Eugene Wigner. When one of Dirac's old friends, who had not yet heard of the marriage, dropped into his home he found with Dirac an attractive woman who served tea and then sat down comfortably on a sofa. "How do you do!" said the friend, wondering who the woman might be. "Oh!" exclaimed Dirac, "I am sorry. I forgot to introduce you. This is ... this is Wigner's sister." '

Magic squares

'The peculiar interest of magic squares ... lies in the fact that they possess the charm of mystery. They appear to betray some hidden intelligence which by a preconceived plan produces the impression of intelligent design, a phenomenon which finds its close analogue in nature.'

Paul Carus

Which are the most beautiful?

In 1988 a questionnaire was published in *The Mathematical Intelligencer*, inviting readers, most of whom are professional mathematicians, to give a selection of theorems a score from 1 to 10 for *beauty*. This is, of course, a very crude question, but more than seventy replies were received, and the average score for each theorem calculated.

Here are the first nine theorems in the final ranking, and the last two, together with the average of the scores given to them.

Rank	Theorem	Average
(1)	$e^{i\pi} = -1$	7·7
(2)	Euler's formula for a polyhedron: $V + F = E + 2$	7·5
(3)	The number of primes is infinite.	7·5
(4)	There are 5 regular polyhedra.	7·0

(5) $1 + \frac{1}{2^2} + \frac{1}{3^2} + \frac{1}{4^2} + \ldots = \pi^2/6$ 7.0

(6) A continuous mapping of the closed unit disk into itself has a fixed point. 6.8

(7) There is no rational number whose square is 2. 6.7

(8) π is transcendental. 6.5

(9) Every plane map can be coloured with 4 colours. 6.2

(10) Every prime number of the form $4n + 1$ is the sum of two integral squares in exactly one way. 6.0

(23) The maximum area of a quadrilateral with sides a,b,c,d is $[(s - a)(s - b)(s - c)(s - d)]^{1/2}$, where s is half the perimeter. 3.9

(24) $\frac{5[(1 - x^5)(1 - x^{10})(1 - x^{15}) \ldots]^5}{[(1 - x)(1 - x^2)(1 - x^3)(1 - x^4) \ldots]^6}$ 3.9
$= p(4) + p(9)x + p(14)x^2 + \ldots,$
where $p(n)$ is the number of partitions of n.

There is a postscript to these results. According to G. H. Hardy, writing in 1927,

'. . . if I had to select one formula from all Ramanujan's work, I would agree with Major MacMahon in selecting . . . , viz.

$$p(4) + p(9)x + p(14)x^2 \ldots = \frac{5\{(1 - x^5)(1 - x^{10})(1 - x^{15}) \ldots\}^5}{\{(1 - x)(1 - x^2)(1 - x^3) \ldots\}^6},$$

where $p(n)$ is the number of partitions of n.'

This formula is the one that came *last* in the questionnaire. The respondents, who of course were not told that the formula was due to Ramanujan, found it no more beautiful than a standard result in traditional geometry. How fashions in mathematics change! (Compare Reuben Hersh's opinion, page 29.)

The arrival of Indian numerals

This is the first evidence of the spread of Indian numerals towards the West. It was written by Severus Sebokht, a Nestorian Bishop, in 662:

'I will omit all discussion of the science of the Indians, a people not the same as the Syrians; of their subtle discoveries in astronomy, discoveries that are more ingenious than those of the Greeks and the Babylonians; and of their valuable methods of calculation which surpass description. I wish only to say that this computation is done by means of nine signs. If those who believe, because they speak Greek, that they have arrived at the limits of science, [would read the earlier texts], they would perhaps be convinced, even if a little late in the day, that there are others also who know something of value.'

Three hundred years later, they were first mentioned in a European manuscript, the *Codex Vigilanus*, dating from 976.

'So with computing symbols. We must realize that the Indians had the most penetrating intellect, and other nations were way behind them in the art of computing, in geometry, and in other free [?] sciences. And this is evident from the nine symbols with which they represented every rank of number at every level.'

Surprise

Jacques Ozanam was a correspondent of Leibniz, the author of a famous book of mathematical recreations, and De Moivre's teacher. Here he is writing in his *Cursus Mathematicus* of 1712.

'The greatest part of the Lovers of Mathematics are won to that Science by its sensible Beauties only, they are taken by the Wonders that it works and delighted by its admirable Phaenomena; They are willing to know what they have admir'd, to perform those Things which at first they could not account for; and take pleasure in surprising others, as themselves have been surprised.'

How does mathematics exist?

Does the mathematician create mathematics, or merely discover what previously existed in some real but non-objective world? Here are the views of three *Platonists*, who take the latter view.

'Nor do we give laws to the intellect or to things according to our own judgement, but like faithful scribes, those laws which are borne on the voice of nature itself and proclaimed, we take up and describe.'

Cantor

'No! The mathematician cannot arbitrarily produce something, any more than the geographer; he too can only discover what exists, and give it a name.'

Frege

'I believe that the numbers and functions of analysis are not the arbitrary product of our minds; I believe that they exist outside of us with the same character of necessity as the objects of objective reality; and we find or discover them and study them as do the physicists, chemists and zoologists.'

'If I am not mistaken, there exists a whole world which contains all mathematical truth, to which we have no access except through intelligence, just as there exists a world of physical realities: both independent of us, both of divine creation, which only seem distinct because of our weakness of intellect, which are but one and the same thing for a powerful mind, and the synthesis being partially revealed in this marvellous correspondence between abstract Mathematics on the one hand and Astronomy and all branches of Physics on the other.'

Hermite

The cost of everything

Mathematics has been applied to every science and philosophy under the sun, not excluding the moral sciences. The *Computatio Universalis* (1697) of Samuel Foley claimed to be 'an essay attempting in a geometrical method to demonstrate a universal standard where one may judge the true value of everything in the world . . . The design of the following essay is to put people in a right method of good husbandry, and to assist them to produce as much happiness as is procurable by them, showing them how to compute their time and riches, and to compare them with other things.'

Foley started with six Definitions, five Postulates, and three Axioms. He then inferred, calling in each case upon the appropriate definitions, postulates and axioms, that:

'1st Proposition: If a man do live from his birth to the end of 64 years, that is 64 apparent years of time, his real age is but 32 years.

2nd Proposition: If such a man has an estate of inheritance of £120 per annum, his whole real estate is £4,940 sterling.'

From which he leads up to the 6th Proposition, in which he infers that one minute of such a man's time is worth approximately 3/10 of a farthing, and the 11th Proposition, in which he discovers the value of an indifferent horse, purchased by such a man at the age of 33 years.

A fashion for algebra

'[Algebra] is become so common amongst us, because of its engaging Beauty and vast Use in all parts of the Mathematics, that even Ladies of the highest Quality have been induced to learn it; the Duchess of E . . . has attained so great a degree of Perfection, as well in Numbers as Geometry, that Persons who make the Greatest Figure for Learning have earnestly sought for the Honour of her Conversation.'

Jacques Ozanam

An encomium

' 'Tis by Trigonometry only, that the Causes of the Phaenomena's and Changes which happen in the Universe, can with any certainty, be discovered; which causes depend solely on the Figure and Motion of the World, and its Parts, nor can any one arrive at the knowledge of the Motions of the Coelestial Bodies, but by that of the most simple Figures, which are Triangles, whose Mensuration is taught by Trigonometry. By whose assistance Engineers measure accessible and inaccessible Distances . . . Geographers measure the Distances of places situated on the Surface of the Earth; Astronomers, the Distance of stars, whose Longitude and Latitude are known; the Art of Navigation depends entirely on Trigonometry . . . The Usefulness of Trigonometry is so great, in all parts of the Mathematics, that it is in a manner impossible to live without it, and the most important and useful parts of knowledge would be utterly lost, if Mankind were ignorant of it.'

Jacques Ozanam

Mathematical feeling

The Russian educator V. A. Krutetskii, in his classic work, *The Psychology of Mathematical Abilities in Schoolchildren*, analysed the abilities of outstandingly talented young mathematicians, including their aesthetic feelings:

'This experience of the elegance of a solution was very characteristic of the capable pupils we observed. "A beautiful solution!" "This method, like a good chess combination, evokes a feeling of pleasure in me," the pupils said. And their whole demeanor testified to the aesthetic feeling they were experiencing: their eyes sparkled, they rubbed their hands in satisfaction and smiled, they invited one another to admire a keen train of thought or a particularly "elegant" solution.'

Hungarians solving problems

Why does Hungary have such a brilliant reputation for producing mathematicians? An obvious answer is – look at their schools. Here is one account, by Theodore von Kármán, one of the founders of modern aeronautics.

'[The Minta, or Model Gymnasium] became the model for all Hungarian high schools. Mathematics was taught in terms of everyday statistics. We looked up the production of wheat in Hungary, set up tables, drew graphs, learned about the "rate of change" which brought us to the edge of calculus. At no time did we memorize rules from a book. Instead, we sought to develop them ourselves . . . The Minta was the first school in Hungary to put an end to the stiff relationship between the teacher and the pupil which existed at that time. Students could talk to the teachers outside of class and could discuss matters not strictly concerning school. For the first time in Hungary a teacher might go so far as to shake hands with a pupil in the event of their meeting outside of class.

Each year the high schools awarded a national prize for excellence in mathematics. It was known as the Eötvös Prize. Selected students were kept in a closed room and given difficult mathematics problems, which demanded creative and even daring thinking. The teacher of the pupil who won the prize would gain great distinction, so the competition was keen and teachers worked hard to prepare their best students. I tried out for this prize against students of great attainments, and to my delight I managed to win. Now, I note that more than half of all the famous expatriate Hungarian scientists, and almost all the well-known ones in the United States have won this prize. I think that this kind of contest is vital to our educational system, and I would like to see more such contests encouraged here in the United States and in other countries.'

Studying the teacher

Many children find themselves lost in mathematics lessons, floating on the surface of a sea of incomprehension. How much worse the situation of an African child, taught mathematics from an English-language textbook, based on a strange culture. This is the result.

'No occasion arises for a child to use his talent for discovery, or his curiosity, in relation to the subject matter of a course. He is forced to repeat aloud collections of words that, from his point of view, make no sense. He knows that he must please the teacher in order to survive, but he finds what is taught incomprehensible. Therefore he tries to find other ways to survive; he uses his wit to anticipate the teacher. If the material being taught has no apparent pattern, at least he can figure out the teacher. Often the teacher has come from the same Kpelle background as the student, and so his words and actions are more or less predictable. The teacher's words and actions do not seem haphazard, disorganized, and irrelevant. This leads the child to guess at the best way to please the teacher, using nonacademic, social clues. Even when the teacher is not of Kpelle origin, he behaves in a way that makes sense to the Kpelle child, because many of the patterns of Westernized Liberian life duplicate the authority structures of Kpelle life.

It is common for children to shout out answers to the teacher before he has finished stating a problem. They try to outguess each other to show the teacher how smart they are. For example, a teacher gave the problem, "Six is two times what number?" When the children had heard the words "six", "two", and "times", they shouted "twelve". Their experience with the teacher showed that he was always asking the times table, and they guessed he was asking it again. They were wrong – and probably were never quite sure why. They were told the answer was "three", which probably confirmed their idea that school made no sense whatever.'

The joy of maths

'We have no knowledge, that is, no general principles drawn from the contemplation of particular facts, but what has been built up by pleasure, and exists in us by pleasure alone. The man of science, the chemist, the mathematicians, whatever difficulties and disgusts they may have to struggle with, know and feel this.'

Wordsworth

Newton keeps Sylvester working

James Joseph Sylvester (1814–1897) was one of the greatest English mathematicians, especially notable, with Arthur Cayley (1821–1895), as an algebraist and founder of the theory of invariants.

'Five years later (1864) he contributed to the Royal Society of London what is considered his greatest mathematical achievement. Newton, in his lectures on algebra, which he called "Universal Arithmetic" gave a rule for calculating an inferior limit to the number of imaginary roots in an equation of any degree, but he did not give any demonstration or indication of the process by which he reached it. Many succeeding mathematicians such as

Euler, Waring, Maclaurin, took up the problem of investigating the rule, but they were unable to establish either its truth or inadequacy. Sylvester in the paper quoted established the validity of the rule for algebraic equations as far as the fifth degree inclusive. Next year in a communication to the Mathematical Society of London, he fully established and generalized the rule. "I owed my success," he said, "chiefly to merging the theorem to be proved in one of greater scope and generality. In mathematical research, reversing the axiom of Euclid and controverting the proposition of Hesiod, it is a continual matter of experience, as I have found myself over and over again, that the whole is less than its part." '

Simplicity in mathematical physics

'. . . if one chooses to look into simple problems, one has a bigger chance of coming close to fundamental structures in mathematics. Second, one must have a certain appreciation of the value judgment of mathematics.

Most papers in theoretical physics are produced in the following way: A publishes a paper about his theory. B says he can improve on it. Then C points out that B is wrong, and so forth. Most of the time, it turns out that the original idea of A is totally wrong or irrelevant.

Mathematical theorems are proved, or supposed to be proved. In theoretical physics, we are pursuing instead a guessing game, and guesses are mostly wrong.

It is important to know what other research workers in one's field are thinking about. But to make real progress, one must face original simple physical problems, not other people's guesses.

. . . the simpler the problem, the more the analysis is likely to be close to some basic mathematical structure. I can illustrate this with the following observation: If there is a mathematics-based winning strategy in the game of chess, or in Wei-qi (known in the United States by the later Japanese name of "go"), then it must be in Wei-qi, because Wei-qi is a simpler, more basic game.

Many theoretical physicists are, in some ways, antagonistic to mathematics, or at least have a tendency to downplay the value of mathematics. I do not agree with these attitudes. I have written:

Perhaps because of my father's influence, I appreciate mathematics more. I appreciate the value judgement of the mathematician, and I admire the beauty and power of mathematics: there are ingenuity and intricacy in tactical maneuvers, and breathtaking sweeps in strategic campaigns. And, of course, miracle of miracles, some concepts in mathematics turn out to provide the fundamental structures that govern the physical universe!'

C. N. Yang, who shared the Nobel prize for physics in 1957

Aubrey on Education

John Aubrey (1626–1697), author of *Brief Lives*, had revolutionary educational views, arguing that 'a Schoole should be indeed the house of play and pleasure; and not of feare and bondage', intending that his pupils should take 'so great a delight in their Studies, that they would learn as fast as one could teach them.'

He also refused to start with the learning of Latin, commencing instead with 'Cookery, Chemistry, Cards (They may have a Banke for wine, of the money that is wonne at play every night), Merchants, Accompts, the Mathematicks'. This is his mathematical syllabus for a Young Gentleman.

Years of Age	*Years at School*	
10	1st	Writing and introduction to algorithms and graphics, globes. Show 'em constellations and plain trigonometry. (Let their nurses teach 'em to number about six years old.)
11	2nd	Continuation with arithmetic and begin with literal algebra and geometry, and when they have passed the first form, surveying, continued use of the globes,

		astrolabe, spherical trigonometry. Mr Paschal's *Scheme* for Greek and Latin, and things in the Real Character.
12	3rd	Kersey's *Algebra*, together with Brancker's and Dr Pell's *Questions* and Mr Oughtred's *Clavis*. And a way of working, daily practising solving of questions. Practical mathematics; for example, perspective, architecture, fortifications, tactics.
13 14	4th	Still keeping up and continuing with algebraical exercises, but now beginning to improve as to the solution of questions or cases in the civil law. And so to proceed in that method to the common law, and to the solution of cases of conscience.
15 16 17 18		*Idem*

Appearances

'In the dusk of a lane I met a shepherd with his sheep. A small dog with the expression of a professor of mathematics does all the work.'

H. V. Morton

The *Ladies' Diary*, or, *Woman's Almanac* (1704–1841)

The *Ladies' Diary* was first published in 1704 and consisted initially of recipes, sketches of notable women, and articles on education and health, naturally appealing to its readership.

Within a short time, however, its contents changed, to be replaced by rebuses, enigmas and mathematical questions. That it not only survived but flourished is a blow in the eye to those who suppose that women cannot be interested in mathematics, and proof that caricatures of women as mathematically incapable were less well-established in the early eighteenth century than the late twentieth. Although men soon proposed and answered many of the questions, women continued to contribute as posers and solvers, often under pseudonyms, such as 'Anne Philomathes'.

The cover of the 1738 issue described it as 'Containing many Delightful and Entertaining Particulars, Peculiarly Adapted for the use and Diversion of the Fair-Sex', and promised women readers that the cultivation of their minds would add to their attractiveness: 'Wit join'd to Beauty . . . leads more Captive than the Conqu'ring Sword.' The first editor wrote that, 'foreigners would be amaz'd when I show them no less than 4 or 500 several letters from so many several women, with solutions geometrical, arithmetical, algebraical, astronomical, and philosophical.'

In subsequent years, many of the questions were very difficult and were answered by almost all the famous mathematicians of the eighteenth century. When it ceased publication, this role was taken over by the mathematical problem column of *The Educational Times*. Great mathematicians such as Charles Hermite and G. H. Hardy not only contributed problems and solutions, but their contributions are reprinted in their Collected Works.

The stimulus of mathematics

'I once asked a medical researcher how she became a scientist. Through a childhood love of mathematics, she said, and began to speak of the great beauty of mathematical order and regularity . . . As she spoke, the researcher's ordinarily placid face became radiant. I saw that it was the thought of solving the problem that so excited her. It was thinking about thinking that made her glow. Mathematics itself, the first instrument of such intense mental excitement for her, was inextricably bound up with the delight she took in the existence of her own expressive self.'

Vivian Gornick

Hamilton's poetical inspiration

'Augustus de Morgan (1806–1871) was the first Professor of Mathematics at University College, London, a brilliant teacher, and notable eccentric. He wrote: "The moving power of mathematical *invention* is not reasoning but imagination. We no longer apply the homely term *maker* in literal translation of *poet*; but discoverers of all kinds, whatever may be their lines, are makers, or, as we now say, have the creative genius." Hamilton spoke of the *Mécanique analytique* of Lagrange as a "scientific poem"; Hamilton himself was styled the Irish Lagrange. Engineers venerate Rankine, electricians venerate Maxwell; both were scientific discoverers and likewise poets, that is, amateur poets. The proximate cause of [Hamilton's latest] shower of verses was that Hamilton had fallen in love for the second time.

Wordsworth advised him to concentrate his powers on science; and, not long after, wrote him as follows: "You send me showers of verses which I receive with much pleasure, as do we all: yet have we fears that this employment may seduce you from the path of science which you seem destined to tread with so much honor to yourself and profit to others. Again and again I must repeat that the composition of verse is infinitely more of an art than men are prepared to believe, and absolute success in it depends upon innumerable *minutiae* which it grieves me you should stoop to acquire a knowledge of . . . Again I do venture to submit to your consideration, whether the poetical parts of your nature would not find a field more favourable to their exercise in the regions of prose; not because these regions are humbler, but because they may be gracefully and profitably trod, with footsteps less careful and in measures less elaborate." '

Clifford's powers of visualization

'The Reverend Percival Frost [author of the classic *Curve Tracing*] once boasted to his brother A. H. Frost (who had done missionary work in India) of the remarkable space perception possessed by the young William Kingdon Clifford [1845–1879]. Now the brother had brought back with him from India a complicated three-dimensional puzzle in the form of a sphere composed of a number of cleverly interlocked pieces, and the challenge of the puzzle was to take it apart. A.H. could scarcely believe the incredible things Percival said about the boy Clifford, and so he asked Percival to invite the youngster to tea and see if he could unlock the three-dimensional puzzle. Forthwith the boy was invited and shown the puzzle. Without touching the puzzle, Clifford carefully looked it over for a few minutes, and then sat with his head in his hands for a few more minutes. He then picked up the puzzle and, to A.H.'s astonishment, immediately took it apart.'

The real and the fancy

'The spirit of Plato dies hard. We have been unable to escape the philosophical tradition that what we can see and measure in the world is merely the superficial and imperfect representation of an underlying reality. Much of the fascination of statistics lies embedded in our gut feeling – and never trust a gut feeling – that abstract measures summarising large tables of data must express something more real and fundamental than the data themselves. (Much professional training in statistics involves a conscious effort to counteract this gut feeling.)'

Stephen Jay Gould

Mathematics and the imagination

'There is an astonishing imagination, even in the science of mathematics . . . there was far more imagination in the head of Archimedes than in that of Homer.'

Voltaire

'The mathematician's best work is art, a high and perfect art, as daring as the most secret dreams of imagination, clear and limpid.

Mathematical genius and artistic genius touch each other.'

G. Mittag-Leffler

'You know, for a mathematician he did not have enough imagination. But he has become a poet and now he is doing fine . . .'

Hilbert, about a former student

Leonardo on flight

'See how the beating of its wings against the air supports a heavy eagle in the highly rarefied air; observe also how the air in motion fills the swelling sails and drives a heavily-laden ship. It is clear from these instances, [Leonardo] declares, that a man with wings large enough, and duly attached to his body, might learn to overcome the gravitational pull and the resistance of the air, and conquer the atmosphere.

But suppose the wings break, or the machine turns on its edge and loses controllability: how could the man be saved? asks Leonardo. And he provides four answers, which are today the *A*, the *B*, the *C* and the *D* of flight. In the first place, if the machine turns on its edge at a high altitude, it will have enough time to regain a normal flight condition, provided that its structure is sufficiently strong and its air resistance high. Secondly, all who fly should wear special bags strung together like a rosary and fixed on their backs. In this way, a man falling from a height may avoid hurting himself. Thirdly, accidents of the kind we are discussing may be

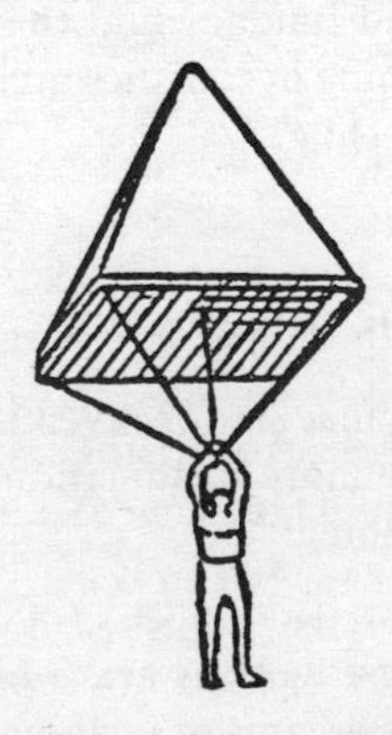

caused by winds and changes in the atmosphere, therefore 'we need a clock to show the hours, minutes and seconds to measure what distance per hour one travels with the course of the wind, and to know the quality and density of the air, and when it has to rain'. Fourthly, a *body moving in the motionless air experiences as much air resistance as is experienced by the same body in a motionless state but exposed to an air motion with the same speed* (Leonardo da Vinci's Principle); if, therefore, we attach to a falling man a horizontal sail, his descent will be resisted by the air; still safer, if the sail is made in the shape shown in the figure. The explanatory note accompanying the drawing says ... If a man has a tent of linen without apertures, twelve ells across and twelve in depth, he can throw himself down from any great height without injury. And so the idea of the modern parachute was born.'

G. A. Tokaty

Chasles is taken for a ride

Michel Chasles (1793–1880) was a brilliant French geometer who also wrote one of the first and best histories of geometry.

'Chasles was an especially ardent French patriot, and his nationalistic pride led to a debacle. When shown some letters, purportedly written by Pascal, in which the laws of gravitational attraction were set out, he eagerly bought them; here was proof of France's priority to Newton's England! A scholar and bibliophile with a comfortable income, Chasles continued to buy documents from one Vrain-Denis Lucas during the period 1861–69.

The details of his purchases seem incredible. He bought over 27,000 letters, for about 140,000 francs. There were 175 letters from Pascal to Newton, 139 from Pascal to Galileo, and large numbers written by Galileo. But Lucas provided ancient, non-mathematical letters as well. Included in Chasles' purchases were six from Alexander the Great to Aristotle, one from Cleopatra to Caesar, one from Mary Magdalene to Lazarus, and one from Lazarus to St Peter. Every letter was written on paper, and in French! It is probably true that Chasles, in his ardor and enthusiasm, did not look at many of his 27,000 purchases.

When Chasles disclosed to the French Academy of Sciences

his theory of Pascal's priority to Newton, there was considerable scepticism. Chasles displayed some of his letters, and it was pointed out that the handwriting was not the same as that of letters which were indubitably Pascal's. Various anachronisms appeared. Each was met by a new letter furnished by Lucas, in which the difficulties were explained away. But after several years of controversy, Chasles had to acknowledge defeat. He exhibited his entire stock of 27,000 forged letters, and Lucas was sent to prison for two years.'

Horner's Method

William George Horner (1786–1837) became the most famous of mathematical schoolmasters, by discovering the method named after him, for the rapid and easy practical calculation of the roots of equations.

The following account is by Augustus de Morgan, who promoted Horner's method and taught it to all his students.

'Wallis happened to choose as his example of the Newtonian method the equation $x^2 - 2x = 5$; from which it has arisen that almost every succeeding writer, who has had anything new to produce in numerical solution, has taken this as one, often as the first, of his instances.

A few years ago, I proposed to one of my classes at University College to carry the solution beyond 32 decimals. Four or five of the members of that class carried the work, independently, to 51 decimal places, or 52 places, the last decimal place being three units in excess, as I now know. The result was published in the article "Involution and Evolution" in the *Supplement* to the *Penny Cyclopaedia*, and in the last edition of my *Arithmetic*.

Previously to the last Christmas vacation, I proposed the same exercise to one of my classes, desiring that the solution might be carried beyond that of their predecessors. Some answers which I received may be described (noting only the *correct* places) as of 75, 65, 63, 58, 57, and 52 decimal places. One answer, however, carried the work to 101 decimal places, or 102 places altogether. The producer of this answer is Mr William Harris Johnston, of Dundalk, and of the Excise Office. On receiving this answer, the extent of which went much beyond any thing I had expected, I requested Mr Johnston, by way of verification, to undertake the equation $y^3 - 90y^2 + 2500y - 16000 = 0$ . . . The relation between the roots of these equations is $y = 30 - 10x$, x being the root of the first, and y that of the second. Accordingly, each place of y is the defect from 9 of the following place of x. Mr Johnston's value of y agrees with what it ought to be, from x, to the very last place.

My reason for publishing this account is, to fix a limit which any future proposed method of solving equations to great extent (I do not speak of special methods for a few figures) must pass before it has established itself against Horner's.'

It was soon noted that Horner had been anticipated in part or in whole by publications of 1816, 1807 and 1804. Neither knew that the Chinese were using 'Horner's method' in 100 BC to find the roots of quadratic and cubic equations, and in AD 1247 Ch'in Kiu-Shao extended the method to equations of higher degree.

Problem-solving

'Approach your problem from the right end and begin with the answers. Then one day, perhaps you will find the final question.'

R. van Gulik

'It isn't that they can't see the solution. It is that they can't see the problem.'

G. K. Chesterton

The end of mathematics

'The question of the foundations and the ultimate meaning of mathematics remains open; we do not know in what direction it will find its final solution or even whether a final objective answer can be expected at all. "Mathematizing" may well be a creative activity of man, like language or music, of primary originality, whose historical decisions defy complete objective rationalization.'

Hermann Weyl

Mathematics and Poetry

'No mathematician can be a complete mathematician unless he is also something of a poet.'

Weierstrass

'It is an open secret to the few who know it, but a mystery and stumbling block to the many, that Science and Poetry are own sisters; insomuch that in those branches of scientific enquiry that are most abstract, most formal, and most remote from the grasp of the ordinary sensible imagination, a higher power of imagination akin to the creative insight of the poet is most needed and most fruitful of lasting work.'

Frederick Pollock

'Poetry is an exact a science as geometry.'

Gustave Flaubert

Frederick Pollock was the son of Chief Baron Pollock, who was Senior Wrangler at Cambridge University in 1806 and who wrote his last paper, on the theory of numbers, at the age of 84.

Projection, a novel device

In 1845 T. S. Davies published the first issue of *The Mathematician*, containing a mixture of popular articles and problems. He was especially impressed by the continental geometers, whose methods, he noted, did not follow the English tradition:

'The paper of Brianchon, to which reference has already been made, contains several things which are worthy of especial notice. Its professed object is, to shew the value of the perspective projection in the demonstration of theorems respecting plane figures:– a process occasionally resorted to by geometers of our own country, but generally esteemed (whether justly or not, I am not disposed to discuss *here*) to be a species of "geometrical trickery". It is sufficient to say just now, of such "trickeries", that they furnish unquestionable evidence of the truth of theorems respecting the conic sections when their correlatives for the circle have been proved; and though *our notions of elegance or geometrical purity may be violated by their use*, we are still *placed a step in advance . . .*'

His journal did not long survive. At the end of the third volume, he expressed sentiments that have long been used to characterize the English:

'I cannot take leave of the many friends I have been accustomed to meet in these pages, and from whose writings I have derived much instruction, without a feeling of deep regret for the inevitable discontinuance of this work . . . The science now acquired, or even sought for, is becoming more superficial:– the mere results of investigation, are alone objects of attraction. These results are only valued for their *uses* either in commerce or manufactures; or perchance occasionally for their aid in "getting up a popular and illustrated book for the market". Reasoning the most unsatisfac tory, and assumption the most unwarrantable, replace the careful investigations and rigorous logic of the philosophers of other days . . . As a people we are indeed become so *practical* and so *physical*, that there is now no journal open to the discussion of pure geometrical science:– except, indeed, it present itself occasionally under the guise of "practical utility" or "physical application".'

That beautiful hypothesis

'Laplace presented a copy of an early volume of his *Mécanique celeste* to Napoleon, who studied it very carefully. Sending for Laplace, he said, "You have written a large book about the universe without once mentioning the author of the universe." "Sire," Laplace replied, "I have no need of that hypothesis."

When Lagrange heard of Laplace's reply to Napoleon, he is said to have shaken his head at his colleague's skepticism, commenting, "But it is a beautiful hypothesis just the same. It explains so many things." '

Obsessions

'A man or woman obsessed with sex arouses, rightly, a certain feeling of distaste. We must learn to cultivate a similar reaction in the presence of a mathematician. Both types are excessive, and therefore imperfect.'

J. W. N. Sullivan

Formulas for primes

There is no formula which produces the sequence of primes in order, in a way that could be calculated. However, there are a number of bizarre formulas which produce primes after a fashion.

Thus G. H. Hardy proved that given any integer N, the largest prime factor of N is given by

$$\theta(N) = \lim_{r=\infty} \lim_{m=\infty} \lim_{n=\infty} \sum_{v=0}^{m} [1 - (\cos\{(v!)^r \pi / N\})^{2n}],$$

As Hardy noted, the practical value of this conclusion is *nil*, though its curiosity value is high, and the ratio interest/value is enormous, indeed, infinite, which would have pleased Hardy, who never liked to think that any of his work was of any pragmatic use.

More seriously, there is a remarkable polynomial formula, whose set of positive values is the set of prime numbers as the variables range over the natural numbers. There are actually many such formulas, all the consequences of the solution of Hilbert's Tenth Problem, which was solved over a period of many years by Martin Davis, Julia Robinson, and, finally, Yuri Matijasevič.

The first polynomial discovered with this property had 24 unknowns and was of the 37th degree. Subsequently, much simpler polynomials were discovered. J. P. Jones tried to minimize the number of *symbols* used in such a polynomial. The following polynomial uses only 325 symbols:

$$(k+2)\{1 - [wz+h+j-q]^2 - [(gk+2g+k+1)(h+j)+h-z]^2$$
$$- [2n+p+q+z-e]^2 - [16(k+1)^3(k+2)(n+1)^2+1-f^2]^2$$

$$- [e^3(e+2)(a+1)^2+1-o^2]^2-[(a^2-1)y^2+1-x^2]^2$$
$$- [16r^2y^4\{a^2-1)+1-u^2]^2 - [((a+u^2(u^2-a))^2-1)(n+4dy)^2$$
$$+ 1 - (x-cu)^2]^2 - [n+l+v-y]^2$$
$$- [(a^2-1)l^2+1-m^2]^2 - [ai+k+1-l-i]^2$$
$$- [p+l(a-n-1)+b(2an+2a-n^2-2n-2)-m]^2$$
$$- [q+y(a-p-1)+s(2ap+2a-p^2-2p-2)-x]^2$$
$$- [z+pl(a-p)+t(2ap-p^2-1)-pm]^2\}.$$

Babbage swimming in circles

Charles Babbage, inventor of the Difference Engine, a forerunner of modern computers, was interested in mechanisms as a small child. As he records, 'My invariable question on receiving a new toy, was, "Mamma, what is inside it?" Until this information was obtained those around me had no repose, and the toy itself, I have been told, was generally broken open if the answer did not satisfy my own little ideas of the "fitness of things".'

The following passage describes an adult attempt at mechanical invention.

'One day an idea struck me, that it was possible, by the aid of some simple mechanism, to walk upon the water, or at least to keep in a vertical position, and have head, shoulders, and arms above water.

My plan was to attach to each foot two boards closely connected together by hinges themselves fixed to the sole of the shoe. My theory was, that in lifting up my leg, as in the act of walking, the two boards would close up towards each other; whilst on pushing down my foot, the water would rush between the boards, cause them to open out into a flat surface, and thus offer greater resistance to my sinking in the water.

I took a pair of boots for my experiment, and cutting up a couple of old useless volumes with very thick binding, I fixed the boards by hinges in the way I proposed. I placed some obstacle between the two flaps of each book to prevent them from approaching too nearly to each other so as to impede their opening by the pressure of the water.

I now went down to the river, and thus prepared, walked into the water. I then struck out to swim as usual, and found little difficulty. Only it seemed necessary to keep the feet farther apart. I now tried the grand experiment. For a time, by active exertion of my legs, I kept my head and shoulders above water and sometimes also my arms. I was now floating down the river with the receding tide, sustained in a vertical position with a very slight exertion of force.

But unfortunately one pair of my hinges got out of order, and refused to perform its share of the propulsion. The result was that I became lop-sided. I was therefore obliged to swim, which I now did with considerable exertion; but another difficulty soon occurred, – the instrument on the disabled side refused to do its share in propelling me. The tide was rapidly carrying me down the river; my own exertions alone would have made me revolve in a small circle, consequently I was obliged to swim in a spiral. It was very difficult to calculate the curve I was describing upon the surface of the water, and still more so to know at what point, if at any, I might hope to reach its banks again. I became very much fatigued by my efforts, and endeavoured to relieve myself for a time by resuming the vertical position.

After floating, or rather struggling for some time, my feet at last touched the bottom. With some difficulty and much exertion I now gained the bank, on which I lay down in a state of great exhaustion.

This experiment satisfied me of the danger as well as of the practicability of my plan, and ever after, when in the water, I preferred trusting to my own unassisted powers.'

Charles Babbage

Putting two and two together

Tolstoy, the novelist, lying on his death bed, was urged to return to the embrace of the Russian Orthodox Church. He replied, 'Even in the shadow of death two and two don't make six.'

'During the morning I thought about thought, and decided it would really be easier to believe in the Divinity and Christ than in twice two making four.'

Sylvia Townsend Warner

The Death of Abel, aged 27 years

'The paper [on elliptic functions] was only two brief pages, but of all his many works perhaps the most poignant. He called it only "A Theorem"; it had no introduction, contained no superfluous remarks, no applications. It was a monument resplendent in its simple lines – the main theorem from his Paris memoir, formulated in few words.

Abel fully understood the beauty of his result. He was unaware that it was the last work which fate had granted him leave to produce; in the last few lines he promised to give many applications which would throw new light on analysis.

Legendre wrote to him, concluding, "I urge you to let this new theory appear in print as quickly as you are able. It will be of great honor to you, and will universally be considered the greatest discovery which remained to be made in mathematics. Good-bye, Monsieur, you must be happy over your success, over the content of your works. It is my hope that you will become still more so by finding a position in society which will permit you to devote yourself entirely to your genial inspiration. Your humble servant, Legendre."

The improvement in the state of Abel's health proved to be temporary, and the family at Froland began to understand that, at best, the illness would be of long duration. A certificate from the physician, dated February 21, 1829, was sent to Holmboe, asking him to report on Abel's condition to the Collegium:

"At the request of Herr Docent Abel, and in the capacity of being his physician – since he is himself unable to write – I shall report to the high Academic Collegium that shortly after his arrival at Froland Ironworks he suffered a severe attack of pneumonia with considerable expectoration of blood, which ceased after a brief period. But a chronic cough and great weakness have compelled him to rest in bed, where he must still remain; furthermore, he cannot be permitted to be exposed to the slightest variation in temperature."

Crelle, in Berlin, was informed of the serious situation; he was desperate over the slowness of the authorities which made it impossible for him to bring the help and cheer which he still believed could save his young friend. Crelle had been informed of the intervention of the French mathematicians, and was concerned that a position in Stockholm or Copenhagen might prove more attractive to Abel. He again turned to the minister of education, urging the greatest haste, and emphasizing the smallness of the amount involved and the modesty of Abel's demands.

On April 8, a delighted and hopeful Crelle could write to Abel that his intense request had brought a favorable response:

"Now, my dear precious friend, can I bring you good news. The Education Department has decided to call you to Berlin for an appointment. I heard it this very moment from the gentleman in charge of the case in the department, so there is no doubt.

"In what capacity you will be appointed and how much you will be paid I cannot tell you, for I do not know myself. I only spoke with the gentleman in passing in a large gathering, so that in the moment I heard nothing more. As soon as I obtain further details I shall let you know.

"I only wanted to hurry to let you hear the main news; you may be certain that you are in good hands. For your future you need no longer have any concern; you belong to us and are secure. I am so glad – it is as if the wish had come true for myself. It has cost a good deal of effort, but, God be praised, it succeeded."

At Froland Abel was surrounded by solicitous attention and nursed lovingly by Crelly [his fiancée], assisted by the two oldest daughters in the Smith family, Marie and Hanna. The latter wrote a few recollections of Abel's last hours.

During March it became apparent that the end was near; the weakness and cough increased and he could remain out of bed only the few minutes while it was being made. Occasionally he would attempt to work on his mathematics, but he could no longer write. Sometimes he lived in the past, talking about his poverty and about Fru Hansteen's goodness. Always he was kind and patient.

Crelly's sorrow was so great that she had difficulties containing it in his presence, so Marie or Hanna were always present with her at the sickbed. Niels Henrik suffered from insomnia as the cough became more severe; he was afraid of being left alone, and a nurse was engaged for relief during the night watches. [Hanna wrote:]

"He endured his worst agony during the night of April 5. Toward morning he became more quiet and in the forenoon, at 11 o'clock, he expired his last sigh. My sister and his fiancée were with him in the last moment, and saw his quiet passing into the arms of death." '

The language of Nature

'The great book of Nature lies ever open before our eyes and the true philosophy is written in it . . . But we cannot read it unless we have first learned the language and the characters in which it is written . . . It is written in mathematical language and the characters are triangles, circles, and other geometrical figures . . .'

Galileo

Cardan's death

Girolamo Cardano (1501–1576: his name is often Anglicized to 'Cardan') was a brilliant mathematician, who is most famous however for having published in his *Ars magna* the solution of the cubic equation, which he had obtained from Niccolo Tartaglia (*c.* 1500–1557) only after swearing never to make the secret public. He was also an astrologer, doctor, and the inventor of the differential gearing.

'What shall we say of those four extraordinary gifts Nature had endowed him withal? Which were: 1. That he could fall into an ecstasy whenever he pleased. 2. That he could see whatever he pleased. 3. That he foresaw in his sleep what was to befall him. And, 4. That he could foretell it likewise by certain marks upon his nails . . . He is justly condemned for his impudence in calculating Jesus Christ's nativity. His astrological predictions are said to have been oftentimes warranted by the event; but he owns that in relation to himself his art had deceived him. Some say that Cardan having foretold he should die at a certain time, he abstained from nourishment that by his death he might verify his prediction, and his life might not bring a scandal upon his profession. Few people, in the like case, stand up with so much courage, and vindicate the honour of their profession with so much charity. They take heart; they are neither ashamed nor discomposed.'

Bayle's Dictionary (1710)

Madame du Châtelet

Sans doute vous serez célèbre
Par les grands calculs de l'algèbre,
Où votre esprit est absorbé.
J'oserais m'y livrer moi-même;
Mais hélas! $A + D - B$
N'est pas = à je vous aime.

Voltaire to Madame du Châtelet

Gabrielle-Emilie Le Tonnelier de Breteuil, the Marquise du Châtelet-Lomont (1706–1749) translated Newton's *Principia* into French and so became one of his greatest promoters, against the Continental supporters of Descartes. Voltaire admired her, as well as loved her, but could not resist describing her as 'a great man whose only fault was in being a woman. A woman who translated and explained Newton . . . in one word, a very great man'. Her history indicates otherwise. The manner of her death, and André Maurois' passing reference to her translation, suggests the fragility of a woman's life and reputation, then as now, as well as Maurois' ignorance of science.

'Madame du Châtelet is a remarkable instance of the immortality which illicit love will ensure for a woman, provided that its object be illustrious. She was a Mlle de Breteuil, and like many girls of the period was highly educated. She knew Latin and was fond of the sciences. She had studied mathematics, and translated the *Principia* of Newton, adding an algebraic commentary. Add to this the fact that she was, as Voltaire said, "something of a philosopher and a shepherdess," and that she wrote a treatise on *Happiness*. All these labours would to-day be in sorry oblivion had she not been the mistress of Voltaire.

One evening, after a whole day's work on the history of Louis

XV, Voltaire entered Mme du Châtelet's room without being announced, and there found his mistress and Saint-Lambert on a sofa, "conversing together on matters that were neither poetry nor philosophy". He was furiously insulting, went out, and ordered his horses: he would leave Lunéville that very night. Mme du Châtelet forbade the lackeys to order the horses, and went up to Voltaire's room to pacify him. "What!" he cried. "You want me to believe you after what I saw?" – "No," she said. "I still love you, but for some time you have complained that you were not so strong as you were, and that you couldn't go on . . . It is a great grief to me. I am very far from desiring your demise; your health is very dear to me. You, on your side, have always shown great care for mine. Since you agree that you could no longer take care of it, save to your own detriment, how can you be angry that one of your friends is coming to your aid?" – "Ah, Madame," he said. "As always, you are right. But if things must be so, pray, at least, see that they don't go on before my eyes."

Next day Saint-Lambert himself came to see Voltaire and apologised. "My boy," said Voltaire, "I have forgotten everything, and it is I who am wrong. You are in the happy age when a man loves and takes pleasure. Enjoy those moments – they are too brief" . . .

The old couple came back to Cirey reconciled, and were on the point of returning to Paris when Mme du Châtelet, usually so lively, became anxious. At the age of forty-four, she was pregnant.

Her time of waiting was spent in sojourns in Paris and at Cirey. She strove hard to seem cheerful, but she had melancholy forebodings. She thought she would die in childbed. The birth, however, passed off well. When she felt the first pangs, she scribbled some remarks on Newton. In a letter Voltaire remarked: "Whilst scribbling her Newton this evening Mme. du Châtelet felt uncomfortable. She called her maid, who had just time to stretch out her apron and receive a little girl, who was carried to her cradle."

But things went wrong, and on the sixth day she died. M. du Châtelet, Voltaire, and Saint-Lambert were present, all three lamenting. Voltaire, in the extremity of his grief, went out of doors, slipped, and fell. Saint-Lambert had followed him and picked him up. On recovering his senses, he said: "Oh, my young friend, it was *you* who killed her for me!" '

Mathematical madness

'The pursuit of mathematics is a divine madness of the human spirit, a refuge from the goading urgency of contingent happenings.'

A. N. Whitehead

Whitehead's collaborator, Bertrand Russell, had a slightly different angle on the same theme:

'The true spirit of delight, the exaltation, the sense of being more than man, which is the touchstone of the highest excellence, is to be found in mathematics as surely as in poetry.'

. . . and elsewhere, Russell referred to the 'promethean madness that leads the greatest men to strive to become gods'. Who can he have had in mind?

Checkmate

'In mathematics if I find a new approach to a problem, another mathematician might claim he has a better, more elegant solution. In chess, if anybody claims he is better than I, I can checkmate him!'

Emanuel Laskar, world chess champion (1894–1920)

Henri Fabre observes himself

Fabre was also an acute observer of himself, as these biographical passages from *The Life of the Fly* illustrate.

'When I left the normal school, my stock of mathematics was of the scantiest. How to extract a square root, how to calculate and prove the surface of a sphere: these represented to me the culminating points of the subject. Those terrible logarithms, when I happened to open a table of them, made my head swim, with their columns of figures; actual fright, not unmixed with respect, overwhelmed me on the very threshold of that arithmetical cave. Of algebra I had no knowledge whatever. I had heard the name; and the syllables represented to my poor brain the whole whirling legion of the abstruse.

Besides, I felt no inclination to decipher the alarming hieroglyphics. They made one of those indigestible dishes which we confidently extol without touching them. I greatly preferred a fine line of Virgil, whom I was now beginning to understand; and I should have been surprised indeed had any one told me that, for long years to come, I should be an enthusiastic student of the formidable science. Good fortune procured me my first lesson in algebra, a lesson given and not received, of course.

A young man of about my own age came to me and asked me to teach him algebra. He was preparing for his examination as a civil engineer; and he came to me because, ingenuous youth that he was, he took me for a well of learning. The guileless applicant was very far out in his reckoning.

His request gave me a shock of surprise, which was forthwith repressed on reflection:

"*I* give algebra-lessons?" said I to myself. "It would be madness: I don't know anything about it!" . . .

It was a fine courage that drove me full tilt into a province which I had not yet thought of entering. My twenty-year-old confidence was an incomparable lever.

"Very well," I replied. "Come the day after to-morrow, at five, and we'll begin." . . .

And now we are together, O mysterious tome, whose Arab name breathes a strange mustiness of occult lore and claims kindred with the sciences of almagest and alchemy. What will you show me? Let us turn the leaves at random. Before fixing one's eyes on a definite point in the landscape, it is well to take a summary view of the whole. Page follows swiftly upon page, telling me nothing. A chapter catches my attention in the middle of the volume; it is headed, *Newton's Binomial Theorem*.

The title allures me. What can a binomial theorem be, especially one whose author is Newton, the great English mathematician who weighed the worlds? What has the mechanism of the sky to do with this? Let us read and seek for enlightenment. With my elbows on the table and my thumbs behind my ears, I concentrate all my attention.

I am seized with astonishment, for I understand! There are a certain number of letters, general symbols which are grouped in all manner of ways, taking their places here, there and elsewhere by turns; there are, as the text tells me, arrangements, permutations and combinations. Pen in hand, I arrange, permute and combine. It is a very diverting exercise, upon my word, a game in which the test of the written result confirms the anticipations of logic and supplements the shortcomings of one's thinking-apparatus.

"It will be plain sailing," said I to myself, "if algebra is no more difficult than this."

I was to recover from the illusion later, when the binomial theorem, that light, crisp biscuit, was followed by heavier and less digestible fare. But, for the moment, I had no foretaste of the future difficulties, of the pitfall in which one becomes more and more entangled the longer one persists in struggling. What a delightful afternoon that was, before my grate, amid my permutations and combinations! By the evening, I had nearly mastered my subject. When the bell rang, at seven, to summon us to the common meal at the principal's table, I went downstairs puffed up with the joys of the newly-initiated neophyte. I was escorted on my way by *a*, *b* and *c*, intertwined in cunning garlands.

Next day, my pupil is there. Black-board and chalk, everything is ready. Not quite so ready is the master. I bravely broach my binomial theorem. My hearer becomes interested in the combinations of letters. Not for a moment does he suspect that I am putting the cart before the horse and beginning where we ought to have finished. I relieve the dryness of my explanations with a few little problems, so may halts at which the mind takes breath awhile and gathers strength for fresh flights.

We try together. Discreetly, so as to leave him the merit of the discovery, I shed a little light upon the path. The solution is found. My pupil triumphs; so do I, but silently, in my inner consciousness, which says:

"You understand, because you succeed in making another understand." '

Fly me to the moon

Hilbert was asked,

'What technological achievement would be the most important? "To catch a fly on the moon." Why? "Because the auxiliary technical problems which would have to be solved for such a result to be achieved imply the solution of almost all the material difficulties of mankind." What mathematical problem was the most important? "The problem of the zeros of the zeta function, not only in mathematics, but absolutely most important!" '

Johanna Gauss

'The beautiful face of a madonna, a mirror of peace of mind and health, tender, somewhat fanciful eyes, a blameless figure – this is one thing; a bright mind and an educated language – this is another; but the quiet, serene, modest and chaste soul of an angel who can do no harm to any creature – that is the best.'

Gauss, describing his bride

The grandeur of Gauss

'It may seem paradoxical, but it is probably nevertheless true that it is precisely the effort after a logical perfection of form which has rendered the writings of Gauss open to the charge of obscurity and unnecessary difficulty. The fact is that there is neither obscurity nor difficulty in his writings, as long as we read them in the submissive spirit in which an intelligent schoolboy is made to read his Euclid. Every assertion that is made is fully proved, and the assertions succeed one another in a perfectly just analogical order; there [being] nothing so far of which we can complain. But when we have finished the perusal, we soon begin to feel that our work is but begun, that we are still standing on the threshold of the temple, and that there is a secret which lies behind the veil and is as yet concealed from us. No vestige appears of the process by which the result itself was obtained, perhaps not even a trace of the considerations which suggested the successive steps of the demonstration. Gauss says more than once that for brevity, he gives only the synthesis, and suppresses the analysis of his propositions. *Pauca sed matura* – few but well-matured – were the words with which he delighted to describe the character which he endeavored to impress upon his mathematical writings. If, on the other hand, we turn to a memoir of Euler's, there is a sort of free and luxuriant gracefulness about the whole performance, which tells of the quiet pleasure which Euler must have taken in each step of his work; but we are conscious nevertheless that we are at an immense distance from the severe grandeur of design which is characteristic of all Gauss's greater efforts.'

Galois' last letter

Evariste Galois (1811–1932) was killed in a duel at the age of twenty. The night before, expecting his own death, he wrote to his friend Auguste Chevalier, summarizing the discoveries he had made concerning groups and the solution of equations by radicals. This is the beginning and the end of that letter.

'My dear friend,

I have made some new discoveries in analysis.

Some are concerned with the theory of equations; others with integral functions.

In the theory of equations, I have sought to discover the conditions under which equations are solvable by radicals, and this has given me the opportunity to study the theory and to describe all possible transformations on an equation even when it is not solvable by radicals.

It will be possible to make three memoirs of all this.

The first is written, and, despite what Poisson has said of it, I am keeping it, with the corrections I have made.

The second contains some interesting applications of the theory of equations. The following is a summary of the most important of these:

You know, my dear Auguste, that these subjects are not the only ones I have explored. My reflections, for some time, have been directed principally to the application of the theory of ambiguity to transcendental analysis. It is desired to see *a priori* in a relation among quantities or transcendental functions, what transformations one may make, what quantities one may substitute for the given quantities, without the relation ceasing to be valid. This enables us to recognize at once the impossibility of many expressions which we might seek. But I have no time, and my ideas are not developed in this field, which is immense.

Print this letter in the *Revue encyclopédique*.

I have often in my life ventured to advance propositions of which I was uncertain; but all that I have written here has been in my head nearly a year, and it is too much to my interest not to deceive myself that I have been suspected of announcing theorems of which I had not the complete demonstration.

Ask Jacobi or Gauss publicly to give their opinion, not as to the truth, but as to the importance of the theorems.

Subsequently there will be, I hope, some people who will find it to their profit to decipher all this mess.

Je t'embrasse avec effusion.

E. Galois.

May 29, 1832.'

The subjection of infinity

The calculus is so readily learnt today that it is easy to overlook the problems that it posed for the eighteenth century, let alone earlier thinkers, mostly to do with quantities infinitely small. Voltaire, on this as other topics, was an acute optimist.

'It is to this method of subjecting everywhere infinity to algebraical calculations, that the name is given of differential calculations or of fluxions and integral calculation. It is the art of numbering and measuring exactly a thing whose existence cannot be conceived.

And, indeed, would you not imagine that a man laughed at you who should declare that there are lines infinitely great which form an angle infinitely little?

That a right [straight] line, which is a right line so long as it is finite, by changing infinitely little its direction, becomes an infinite curve; and that a curve may become infinitely less than another curve?

That there are infinite squares, infinite cubes, and infinites of infinites, all greater than one another, and the last but one of which is nothing in comparison of the last?

All these things, which at first appear to be the utmost excess of frenzy, are in reality an effort of the subtlety and extent of the human mind, and the art of finding truths which till then had been unknown.'

Voltaire (1694–1778)

The role of analogies

'Mathematics is the art of giving the same name to different things.'

'What we must aim at is not so much to ascertain resemblances and differences, as to discover similarities hidden under apparent discrepancies.'

Poincaré

'Difference in similarity and similarity in difference has been called the matter of our science.'

Sylvester

'For above all I love analogies, my most faithful teachers, acquainted with all the secrets of Nature. But, since then, imagination in Science has gone somewhat out of fashion: and, in fear of losing caste, the scientist has grown proverbially matter of fact.'

Kepler

Ulam's working methods

'Ulam . . . is almost exclusively a talking man, a verbal person. When not thinking, which he prefers to do while absently laying out solitaire card games on his bed, what he enjoys most is to talk, to discuss, to argue, to converse, with friends and colleagues. Relying on his phenomenal memory, he carries everything in his head. Or almost, . . . for his pockets, his desk, his briefcase are filled with scrawny, pythic, undecipherable scribblings written on tiny scraps of paper, which are then folded. Disconnected, unrelated, random, they are incomprehensible for anyone else. Once written, these little fragments just pile up, their contents securely filed in his memory. They have served their mnemonic purpose. I have seldom seen him refer to any of them, except during a lecture. Woe just the same to anyone who suggests throwing them away!

The physical act of taking pen to paper has always been painful for him. His mind and his eyes are the obstacles. His mind, because it works so much faster than his fingers that they cannot follow and begin to wriggle like the stylus of a seismograph – hence the jagged, compressed, angular appearance of his handwriting; his eyes, because one is very myopic, the other very presbyotic. Though he claims that by using them separately his vision is better than normal, it is obvious that he has great difficulty focusing at an intermediate range. From childhood fears, then from youthful vanity he spurned wearing glasses, until very recently. Thus Ulam has always had a very hard time bringing himself to write anything for publication, either in long hand or with a typewriter. Machines and other mechanical objects have always turned him off. He is barely able to find and press the record button on a tape recorder. How then does he ever produce a written text? Mainly by talking . . .'

The talent of youth

'Evidence for this is furnished by the fact that the young may be geometers and mathematicians and be wise in such subjects, but they are not thought to be prudent. The reason is that practical wisdom is concerned with particular facts as well, which become known as the result of experience, while a young man cannot be experienced . . . One might indeed inquire further why a boy may be a good mathematician, but cannot be a philosopher or a physicist, May we not say that it is because the subjects of mathematics are reached by means of abstraction, while the principles of philosophy and physics come from experience; and the young have no conviction of the latter . . .'

Aristotle

An extraordinary mind

Stanislav Ulam was one of the most unusual mathematical, and scientific, minds of the century. He gave an insight into his life in his autobiography, *Adventures of a Mathematician*. This passage is taken from an article by his long-time friend and collaborator, Gian-Carlo Rota.

'One morning in 1946, in Los Angeles, Stanislaw Ulam, a newly appointed professor at the University of Southern California, awoke to find himself unable to speak. A few hours later, he underwent a dangerous surgical operation after the diagnosis of encephalitis. His skull was sawed open and his brain tissue was sprayed with antibiotics. After a short convalescence, he managed to recover apparently unscathed.

In time, however, some changes in his personality became obvious to those who knew him. Paul Stein, one of his collaborators at the Los Alamos Laboratory (where Stan Ulam worked most of his life), remarked that, while before his operation Stan had been a meticulous dresser, a dandy of sorts, afterwards he became visibly sloppy in the details of his attire, even though he would still carefully and expensively select every item of clothing he wore.

Soon after I met him in 1963, several years after the event, I could not help noticing that his trains of thought were not those of a normal person, even for a mathematician. In his conversation he was livelier and wittier than anyone I had ever met, and his ideas, which he spouted out at odd intervals, were fascinating beyond anything I have witnessed before or since. However, he seemed to studiously avoid going into any details. He would dwell on any subject no longer than a few minutes, then impatiently move on to something entirely unrelated.

Out of curiosity, I asked John Oxtoby, Stan's collaborator in the thirties (and, like Stan, a former Junior Fellow at Harvard), about their working habits before his operation. Surprisingly, Oxtoby described how at Harvard they would sit for hours on end, day after day, in front of the blackboard. Since the time I met him, Stan never did anything of the sort. He would perform a calculation (even the simplest) only when he had absolutely no other way out. I remember watching him at the blackboard, trying to solve a quadratic equation. He furrowed his brow in rapt absorption, while scribbling formulas in his tiny handwriting. When he finally got the answer, he turned around and said with relief: "I feel I have done my work for the day."

The Germans have aptly called *Sitzfleisch* the ability to spend endless hours at a desk, doing gruesome work. *Sitzfleisch* is considered by mathematicians to be a better gauge of success than any of the attractive definitions of talent with which psychologists regale us from time to time. Stan Ulam, however, was able to get by without any *Sitzfleisch* whatsoever. After his bout with encephalitis, he came to lean on his unimpaired imagination for his ideas, and on the *Sitzfleisch* of others for technical support. The beauty of his insights and the promise of his proposals kept him amply supplied with young collaborators, willing to lend (and risking to waste) their time.

A crippling technical weakness coupled with an extraordinarily creative imagination is the drama of Stan Ulam. Soon after I met him, I was made to understand that, as far as our conversations went, his drama would be one of the Forbidden Topics. Perhaps he discussed it with his daughter Claire, the only person with whom he would occasionally have brutally frank verbal exchanges; certainly not with anyone else. But he knew I knew, and I knew he knew I knew.

Ulam came back to Los Alamos haunted by the fear that his illness might have irreparably damaged his brain. He knew his way of thinking had never been that of an ordinary mathematician, and now less than ever. He also feared that whatever was left of his talents might quickly fade. He decided the time had come for him to engage in some substantial project that would be a fair test for his abilities, and to which his name might perhaps remain associated.'

Education and mathematics compared

'It has well been said that the highest aim in education is analogous to the highest aim in mathematics, namely, to obtain not *results* but *powers*, not particular solutions, but the means by which endless solutions may be wrought.'

George Eliot

Retrograde analysis

A typical chess position can be reached in many different ways, yet chess problemists recognize a type of puzzle in which the solution requires the reconstruction of a sequence of moves, which – it may be inferred from clues hidden in the position – is the *only* sequence which will legally create the position displayed.

W. W. Sawyer here applies the same Sherlock Holmes mentality to an examination problem.

'Occasionally palaeontologists dig up a small fossilized bone and proceed to reconstruct the shape of an extinct animal. A similar activity is possible in regard to examiners, the questions set taking the place of the fossil bone.

A good examination question is not just a shapeless affair; it should contain some interesting design or some surprising result. Such questions are by no means easy to make up. Accordingly, an examiner who is doing research work will usually be on the look-out for some result that he can use in an examination question. Often, in fairly advanced work, some small piece of algebraic manipulation occurs, which can be detached from its context and set as a problem.

For example, some years ago students brought me the following question, which had been set in an examination paper, and which they found hard to solve, or at any rate to solve in a satisfying manner.

Prove that, if[1]

$$\frac{ac - b^2}{a - 2b + c} \qquad \frac{bd - c^2}{b - 2c + d}$$

then the fractions just given are both equal to

$$\frac{ad - bc}{a - b - c + d}.$$

This question has a very definite form, and obviously to hammer it out by a lengthy and shapeless calculation, while verifying the result, would bring one no nearer to the heart of the question. What interested me most was the question, how did the examiner come to think of this question?

The pattern of the question includes the following aspects. $ac - b^2 = 0$ is the condition for the three quantities a, b, c to be in geometrical progression. The numerator of the first fraction contains $ac - b^2$. A similar expression occurs in the numerator of the second fraction. Down below we have expressions $a - 2b + c$ and $b - 2c + d$ which are associated with arithmetical progressions, $a - 2b + c = 0$ being the condition for a, b, c to be in A.P. Again there is a kind of rule by which the denominators could be derived from the numerators; in the third fraction, for example, we have a and d multiplied on top, added below, i.e. the numerator contains ad, the denominator $a + d$. The negative terms are similarly related; on top we have $-bc$, down below $-(b + c)$. This rule applies equally well to the first two fractions; in the first fraction, for instance, $-b^2$ is $-bb$, and we find $-(b + b)$, that is, $-2b$, below it.

To invent a problem so knit together is almost impossible. One does not invent such things; one stumbles upon them. I was certain that the examiner had been *finding the condition for something*, and these fractions had arisen in the course of the work.

The way to begin the problem was fairly obvious, to bring in a new symbol, k, for the value of the fractions. The problem then can be stated as follows.

If $$\frac{ac - b^2}{a - 2b + c} = k \ldots \text{(I)}$$

and $$\frac{bd - c^2}{b - 2c + d} = k \ldots \text{(II)}$$

prove $$\frac{ad - bc}{a - b - c + d} = k \ldots \text{(III)}$$

To bring in such a symbol k is routine procedure, when dealing with the equality of several fractions. (See, for example, Hall and Knight, *Higher Algebra*, Chapter 1.)

What to do next was not at all obvious to me. I tried various methods which, though they led to proofs, did not satisfy me. I continued to think about this question in odd moments, and about a week later, hit on the following approach. Equation (I) can be put in the form $ac - k(a + c) = b^2 - 2bk$. Both sides of this equation are now crying out for an extra term k^2 to complete their pattern. This will "complete the square" on the right-hand side, giving $(b - k)^2$, and give $(a - k)(c - k)$ on the left. So $(a - k)(c - k) = (b - k)^2$. That is to say, *equation* (I) *expresses the fact that* $a - k$, $b - k$, $c - k$ *are in G.P.*

Now we have the whole thing. Equation (II) shows that $b - k$, $c - k$, $d - k$ are in G.P. So $a - k$, $b - k$, $c - k$, $d - k$ are in geometrical progression. But if we multiply together the first and fourth terms of a G.P. the result equals the product of the second and third terms. (Let the G.P. be A, AR, AR^2, AR^3. Then $A \times AR^3 = AR + AR^2$.) So we have

$$(a - k)(d - k) = (b - k)(c - k).$$

If we multiply this out, cancel k^2 and solve the resulting linear equation for k, equation (III) results.

Our conclusions therefore is that the examiner's researches had led him on some occasion to pose the question, "What is the condition that four numbers a, b, c, d must satisfy, if, by subtracting the same number from each of them, a geometrical progression can be obtained?"

The moral of this is not confined to examination questions. It is meant to support the thesis, *where there is pattern there is significance*. If in mathematical work of any kind we find a certain striking pattern recurs, it is always suggested that we should investigate *why* it occurs. It is bound to have some meaning, which we can grasp as an idea rather than as a collection of symbols. It is extremely unsatisfactory to discover a theorem, and only be able to prove it by shapeless calculations. It means that we do not understand what we have discovered.

1. It is assumed that b and c are unequal. The text does not discuss this point, as it is not relevant to the main theme. What suggested the question to the examiner?'

Plato versus Greek ingenuity

'Eudoxus and Archytas had been the first originators of this far famed and highly prized art of mechanics which they employed as an elegant illustration of geometrical truths, and as a means of sustaining experimentally to the satisfaction of the senses, conclusions too intricate for proofs by words or diagrams. As, for example, to solve the problem, so often required in constructing geometrical figures, given two extremes, to find two mean lines of proportion: both these mathematicians had recourse to the aid of instruments adapting to their purposes certain curves, and sections of lines. But what with Plato's indignation at it, and his invectives against it as the mere corruption and annihilation of the good of geometry, which was thus turning its back on the unembodied objects of pure intelligence, to recur to sensation, and ask help (not obtained without base subservience and deprivation) from matter, so it was that mechanics came to be separated from geometry and repudiated and neglected by philosophers, and took its place as a military art.'

Plutarch

Pascal on geometry

'If a demonstration is to carry conviction, each of the steps by which it proceeds must be fully understood; and I can scarcely make this plainer than by describing the steps in a geometrical demonstration . . . for what is beyond the resources of geometry is beyond the reach of man . . . I have chosen geometry as the best method for my purposes, because it is the only art or science which applies the true rules of reasoning . . . as we may learn from experience: where two persons of equal capacity are engaged in debate, the one who knows some geometry (other things being equal) will carry the day – and his skill in argument will moreover continue to grow . . . This true method, if it could be applied, would enable us to construct demonstrations of a very exceptional value, for it would in the main be governed by two rules; it would use no expression of which the sense has not been clearly explained in advance; and it would never offer a proposition which could not be demonstrated from truths already known . . .

Whence it appears that definitions of this [geometrical] type are entirely arbitrary and ought never to be called in question . . .

This is the method of geometry. It neither defines nor proves all the things that are, and for this reason it fails to convince, and yields pride of place to other sciences; but it assumes nothing but what is clear and constant and according to the light of nature, and for this reason it is perfectly reliable . . . It may be thought strange that mathematics should not be able to define any of those things with which she is mainly concerned. She can offer no definition of movement, numbers, space, yet these three things are her particular study . . . But nobody will be surprised if we remark that since this wonderful science is concerned only with the most simple ideas, the very quality which makes them suitable for study also makes them incapable of being defined; so that the absence of a definition is a perfection rather than a defect.'

From primary school to the stars

'Lucy, dear child, mind your arithmetic. You know, in the first sum of yours I ever saw, there was a mistake. You had carried two (as a cab is licensed to do) and you ought, dear Lucy, to have carried but one. Is this a trifle? What would life be without arithmetic but a scene of horrors?'

Sydney Smith

Are carries merely a trifle? There is one very important theorem about carries, due to Kummer.

'The greatest power of a prime p which divides the binomial coefficient $\binom{a+b}{b}$

is p^c, where c is the number of carries needed when adding a and b written to base p.'

The binomial coefficient $\binom{a+b}{b}$ represents the number of ways of selecting b objects from a set of $a + b$ objects, without taking order into account. This theorem about carries turned out to be crucial in the solution of Hilbert's Tenth Problem (page 30).

Voltaire on Newton's achievement

François-Marie Arouet (1694–1778), who called himself Voltaire, was an admirer of Newton, and the lover of the Marquise du Châtelet, who translated Newton's *Principia* into English (page 152). This passage, from his *Letters on the English*, is one of only two sources for the – probably apocryphal – story of Newton and the apple.

'Having by these and several other arguments destroyed the Cartesian vortices, he despaired of ever being able to discover whether there is a secret principle in nature which, at the same time, is the cause of the motion of all celestial bodies, and that of gravity on the earth. But being retired in 1666, upon account of the Plague, to a solitude near Cambridge; as he was walking one day in his garden, and saw some fruits fall from a tree, he fell into a profound meditation on that gravity, the cause of which had so long been sought, but in vain, by all the philosophers, whilst the vulgar think there is nothing mysterious in it. He said to himself, that from what height soever in our hemisphere, those bodies might descend, their fall would certainly be in the progression discovered by Galileo; and the spaces they run through would be as the square of the times. Why may not this power which causes heavy bodies to descend, and is the same without any sensible diminution at the remotest distance from the centre of the earth, or on the summits of the highest mountains, why, said Sir Isaac, may not this power extend as high as the moon? And in case its influence reaches so far, is it not very probable that this power retains it in its orbit, and determines its motion? But in case the moon obeys this principle (whatever it be) may we not conclude very naturally that the rest of the planets are equally subject to it? In case this power exists (which besides is proved) it must increase in an inverse ratio of the squares of the distances. All, therefore, that remains is, to examine how far a heavy body, which should fall upon the earth from a moderate height, would go; and how far in the same time, a body which should fall from the orbit of the moon, would descend. To find this, nothing is wanted but the measure of the earth, and the distance of the moon from it.

Thus Sir Isaac Newton reasoned.'

Hardy's insurance

G. H. Hardy was about to return from Denmark to England, by boat, in appalling weather. So he sent a postcard ahead to announce to the world that 'I have proved Riemann's Hypothesis', which was then as now the Holy Grail of professional mathematicians. Hardy reasoned that God (in whom Hardy did not profess to believe) would not allow the boat to sink, thereby leaving open the suspicion that Hardy had achieved this remarkable feat.

Hilbert and existence

' "Existential" ideas permeated Hilbert's thinking, not only in mathematics, but also in everyday life. This is illustrated by an incident which Helmut Hasse observed at this time. The Society of German Scientists and Physicians was holding its first meeting after the war in Leipzig. In the evenings at the Burgkeller there was much questioning of the type, "What about Professor K. from A., is he still alive?" The 24-year-old Hasse was seated with other young mathematicians at a table quite near to the table shared by Hilbert and his party.

"I heard him put exactly this type of question to a Hungarian mathematician about another Hungarian mathematician. The former began to answer, 'Yes, he teaches at – and concerns himself with the theory of –, he was married a few years ago, there are three children, the oldest . . .' But after the first few words Hilbert began to interrupt, 'Yes, but . . .' When he finally succeeded in stopping the flow of information, he continued, 'Yes, but all of that I don't want to know. I have asked only *Does he still exist?*' " '

The Egyptians teach their children

'In that country arithmetical games have been invented for the use of mere children, which they learn as a pleasure and amusement. They have to distribute apples and garlands, using the same number sometimes for a larger and sometimes for a lesser number of persons: and they arrange pugilists and wrestlers as they pair together by lot or remain over, and show how their turns come in natural order. Another mode of amusing them is to distribute vessels, sometimes of gold, brass, silver, and the like, intermixed with one another, sometimes of one metal only; as I was saying they adapt to their amusement the number in common use, and in this way make more intelligible to their pupils the arrangements and movements of armies and expeditions, and in the management of a household they make people more useful to themselves, and more wide awake; and again in measurements of things which have length, and breadth, and depth, they free us from that natural ignorance of all these things which is so ludicrous and disgraceful.'

Plato

Physics or mathematics?

Freeman Dyson and Harish-Chandra were walking and talking, as mathematicians do, when Harish-Chandra announced, 'I am leaving physics for mathematics, I find physics messy, unrigorous, elusive.' Freeman Dyson replied, 'I am leaving mathematics for physics for exactly the same reasons', and that's what they both did.

Mark Kac tried to explain Freeman Dyson's choice: 'Some years ago I asked a very promising student, who majored in mathematics but who had decided to go into physics, what had prompted his decision. His reply was roughly that in mathematics when you discover something you have the feeling that it has always been there. In physics you have a feeling that you are making a real discovery . . . If doing mathematics or science is looked upon as a game, then one might say that in mathematics you compete against yourself or other mathematicians; in physics your adversary is nature and the stakes are higher.'

A small mistake

'Lagrange . . . imagined that he had overcome the difficulty (of the parallel axiom). He went so far as to write a paper, which he took with him to the Institute, and began to read it. But in the first paragraph something struck him which he had not observed: he muttered: "I must think about it again," and put the paper back in his pocket.'

Augustus de Morgan

Euclid, in his *Elements*, supposed that given a straight line and a point not on it, there was a unique line through the point, parallel to the given straight line. This is the 'parallel axiom' which Euclid had the insight to realize was only an assumption, for which he provided no proof. Later generations of mathematicians made many attempts to prove the 'parallel axiom', always without success, until it was eventually realized that the *denial* of the axiom also led to consistent geometries, the non-Euclidean geometries. The existence of these non-Euclidean geometries finally demonstrated that no proof of the 'parallel axiom' is possible.

Dürer's *Melencolia*

Dürer (1471–1528) was a distinguished mathematician as well as a great artist. He studied perspective mathematically in *Unterweysung der Messung mit Zirkel und Richtscheyt* (1525) which also discusses geometry in three dimensions, and the conic sections, for the first time in German. He was the first person to describe the cycloid. He gives just one example of a proof in this book – but it is the very first proof ever published in German.

He also introduced the basic ideas of descriptive geometry, which are usually credited to Gaspard Monge, and he was the first to show how the regular solids could be constructed by starting with their 'nets'.

The illustration shows Dürer's famous engraving *Melencolia*. The keys and the purse, the bat and the seascape are all associated with melancholy and the planet Saturn. The magic square is associated with Jupiter, and the middle entries in the bottom row make 1514, the date of the engraving.

The curious geometrical object on the left hand of the engraving looks at first sight like a cube with two opposite corners sliced off, but isn't. If a solid is constructed which *will* give the view

shown, in perspective, then it turns out that when the solid is resting on one of its triangular faces, the front orthogonal elevation fits perfectly into the form of the magic square, as the first two figures show. The third figure is the plan.

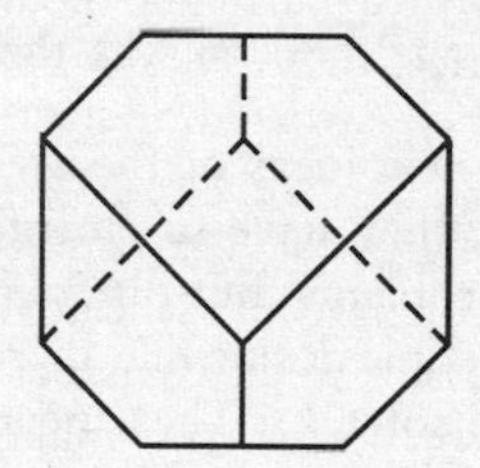
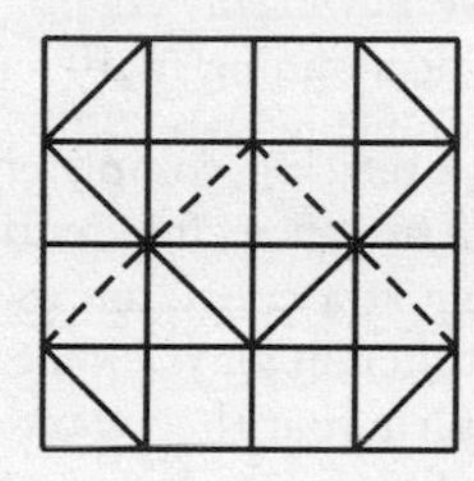
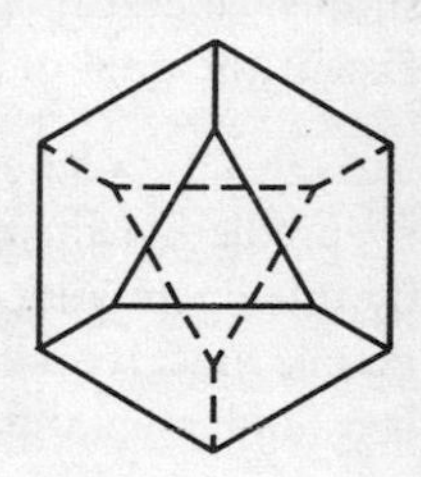

Did Dürer invent this shape? Quite possibly. He had studied polyhedra and he wrote a treatise on perspective.

'It would conform to Dürer's philosophy for it to be his own discovery, for he set great store by originality and wrote in his treatise on human proportions: "For God sometimes granteth unto a man to learn and know how to make a thing whereof, in his day, no other can contrive; and perhaps for a long time none hath been before him and after him another cometh not soon."

As the representation of a polyhedra was seen as one of the main problems of perspective geometry, what better way could Dürer prove his ability in this field, than to include in an engraving a shape that was new and perhaps even unique, and to leave the question of what it was, and where it came from, for other geometricians to solve?'

Michael Atiyah talking

'**Minio:** How do you select a problem to study?
Atiyah: I think that presupposes an answer. I don't think that's the way to work at all. Some people may sit back and say, "I want to solve this problem," and they sit down and say, "How do I solve this problem?" I don't. I just move around in the mathematical waters, thinking about things, being curious, interested, talking to people, stirring up ideas; things emerge and then I follow them up. Or I see something which connects up with something

else I know about, and I try to put them together and things develop. I have practically never started off with any idea of what I'm going to be doing or where it's going to. I'm interested in mathematics; I talk, I learn, I discuss and the interesting questions simply emerge. I have never started off with a particular goal, except the goal of understanding mathematics.

'**Atiyah:** ... You can't develop completely new ideas or theories by predicting them in advance. Inherently, they have to emerge by intelligently looking at a collection of problems. But different people must work in different ways. Some people decide that there is a fundamental problem that they want to solve ... They spend a large part of their life devoted to working towards this end. I've never done that, partly because that requires a single-minded devotion to one topic which is a tremendous gamble.

It also requires a single-minded approach, by direct onslaught, which means you have to be tremendously expert at using technical tools. Now some people are very good at that; I'm not really. My expertise is to skirt the problem, to go round the problem, behind the problem ... and so the problem disappears.'

Sir Christopher Wren and the stocking men

'Sir Christopher Wren proposed to the Silke-Stocking-Weavers of London, viz. a way to weave seven pair or nine paire of stockings at once (it must be an odd Number). He demanded four hundred pounds for his Invention: but the weavers refused it, because they were poor: and besides, they sayd, it would spoile their Trade; perhaps they did not consider the Proverb, That Light Gaines, with quick returnes, make heavy Purses. Sir Christopher was so noble, seeing they would not adventure so much money, He breakes the Modell of the Engine all to pieces, before their faces.'

Love at first sight

Thomas Hobbes, the philosopher, was educated in Latin and Greek, and learnt only the simplest arithmetic as a child. (47 *E. libri I* is a reference to the 47th proposition in the first book of Euclid's *Geometry*, which is the Theorem of Pythagoras.)

'He was 40 years old before he looked on Geometry; which happened accidentally. Being in a Gentleman's library, Euclid's Elements lay open, and 'twas the *47 E. Libri I.* He read the Proposition. *By G–*, says he, (he would now and then sweare an emphaticall Oath by way of emphasis) *this is impossible*! So he reads the Demonstration of it, which referred him back to such a Proposition; which proposition he read. That referred him back to another, which he also read. *Et sic deincips* and so on that at last he was demonstratively convinced of that truth. This made him in love with Geometry.'

Einstein was much younger:

'At the age of 12 I experienced a second wonder of a totally different nature: in a little book dealing with Euclidian plane geometry, which came into my hands at the beginning of a schoolyear. Here were assertions, as for example the intersection of the three altitudes of a triangle in one point, which – though by no means evident – could nevertheless be proved with such certainty that any doubt appeared to be out of the question. This lucidity and certainty made an indescribable impression upon me. That the axiom had to be accepted unproved did not disturb me. In any case it was quite sufficient for me if I could peg proofs upon propositions the validity of which did not seem to me to be dubious. For example I remember that an uncle told me the Pythagorean theorem before the holy geometry booklet had come into my hands. After much effort I succeeded in "proving" this theorem on the basis of the similarity of triangles: in doing so it seemed to me "evident" that the relations of the sides of the right-angled triangles would have to be completely determined by one of the acute angles. Only something which did not in similar fashion seem to be "evident" appeared to me to be in need of any proof at all.'

Bertrand Russell was a similar age:

'At the age of eleven, I began Euclid, with my brother as my tutor. This was one of the great events of my life, as dazzling as first love. I had not imagined that there was anything so delicious in the world. After I had learned the fifth proposition, my brother told me that it was generally considered difficult, but I had found no difficulty whatever. This was the first time it had dawned upon me that I might have some intelligence. From that moment until Whitehead and I finished *Principia Mathematica*, when I was thirty-eight, mathematics was my chief interest, and my chief source of happiness. Like all happiness, however, it was not unalloyed. I had been told that Euclid proved things, and was much disappointed that he started with axioms. At first I refused to accept them unless my brother could offer me some reason for doing so, but he said: "If you don't accept them we cannot go on", and as I wished to go on, I reluctantly admitted them *pro tem*. The doubt as to the premisses of mathematics which I felt at that moment remained with me, and determined the course of my subsequent work.'

Laplace on the course of the universe

'Given for one instant an intelligence which could comprehend all the forces by which nature is animated and the respective positions of the beings which compose it, if moreover, this intelligence were vast enough to submit these data to analysis, it would embrace in the same formula both the movements of the largest bodies in the universe and those of the lightest atom; to it nothing would be uncertain, and the future as the past would be present to its eyes.'

Pierre-Simon Laplace (1749–1827)

Ada Lovelace (1815–1853)

Ada Byron Lovelace was the daughter of Lord Byron and Lady Noel Byron, who had studied algebra, geometry and astronomy with William Frend, notable for his weird opinions of algebra, zero and Newton's theories. In acknowledgement of her mathematical talent, Lord Byron referred to his wife as his Princess of Parallelograms.

Ada never knew the poet, who left her mother soon after she was born. When she was fifteen she first met Charles Babbage, and later became fascinated by his Analytical Engine. When the Italian L. F. Menebrea wrote an account of the Analytical Engine, Ada translated it, and at Babbage's suggestion added extensive and highly original notes of her own.

She died of cancer in 1852, aged 36 years.

Not least of her talents was her broad understanding of the possibilities of Babbage's computer, which compare very favourably indeed with the narrow perspectives of some twentieth century pioneers, who persisted in thinking that a computer was merely for, well, computing.

'[The Engine's] operating mechanism might act upon many other things besides *number*, were objects found whose mutual fundamental relations could be expressed by those of the abstract science of operations, and which should be also susceptible of adaptations to the action of the operating notation and mechanism of the engine. Supposing, for instance, that the fundamental relations of pitched sound in the signs of harmony and of musical composition were susceptible of such expression and adaptations, the engine might compose elaborate and scientific pieces of music of any degree of complexity or extent.'

'The distinctive characteristic of the Analytical Engine, and that which has rendered it possible to endow mechanism with such extensive faculties as bid fair to make this engine the executive right-hand of abstract algebra, is the introduction into it of the principle which Jacquard devised for regulating, by means of punched cards, the most complicated patterns in the fabrication of brocaded stuffs. It is in this that the distinction between the two engines lies. Nothing of the sort exists in the Difference Engine. We may say most aptly, that the Analytical Engine *weaves algebraical patterns* just as the Jacquard-loom weaves flowers and leaves . . . In enabling mechanisms to combine together *general* symbols in successions of unlimited variety and extent, a uniting link is established between the operations of matter and the abstract mental processes of the most abstract branch of mathematical science. A new, a vast, and a powerful language is developed for the future use of analysis, in which to wield its truths so that these may become of more speedy and accurate practical application for the purposes of mankind than the means hitherto in our possession have rendered possible . . .'

Eddington

'I believe that there are 15,747,724,136,275,002,577,605,653,961,181,555,468,044,717,914,527,116,709,366,231,425,076,185,631,031,296 protons in the universe and the same number of electrons.'

The Age of the World

A year for the stake. Three years for the field.
Three lifetimes of the field for the hound.
Three lifetimes of the hound for the horse.
Three lifetimes of the horse for the human being.
Three lifetimes of the human being for the stag.
Three lifetimes of the stag for the ousel.
Three lifetimes of the ousel for the eagle.
Three lifetimes of the eagle for the salmon.
Three lifetimes of the salmon for the yew.
Three lifetimes of the yew for the world
from its beginning to its end, *ut dixit poeta.*

The lives of three wattles, the life of a hound;
The lives of three hounds, the life of a steed;
The lives of three steeds, the life of a man;
The lives of three men, the life of an eagle;
The lives of three eagles, the life of a yew;
The life of a yew, the length of a ridge;
Seven ridges from Creation to Doom.

The flea theme

'An old man insisted on always referring to *mathematical fleas.* He explained that they subtracted from his happiness, divided his attention, added to his misery and multiplied rapidly.'

Jonathan Swift was well aware that the hacks of Grub Street assembled together conspired as much against each other as they did against the public. He had an explanation of this phenomenon, based on Hobbes, Darwin not having yet been born:

Hobbes clearly proves that ev'ry Creature
Lives in a State of War by Nature.
The Greater for the Smallest watch,
But meddle seldom with their Match.
A Whale of moderate Size will draw
A Shole of Herrings down his Maw.
A Fox with Geese his Belly crams;
A Wolf destroys a thousand Lambs.
But search among the rhiming Race,
The Brave are worried by the Base.
If, on *Parnassus*' Top you sit,
You rarely bite, are always bit:
Each Poet of inferior Size
On you shall rail and criticize;
And strive to tear you Limb from Limb,
While others do as much for him.

The Vermin only teaze and pinch
Their Foes superior by an Inch.
So, Nat'ralists observe, a Flea
Hath smaller Fleas that on him prey,
And these have smaller Fleas to bite 'em,
And so proceed *ad infinitum*:
Thus ev'ry Poet in his Kind,
Is bit by him that comes behind;
Who, tho' too little to be seen,
Can teaze, and gall, and give the Spleen . . .

Augustus de Morgan had an expanded nineteenth-century conception of infinity:

Great fleas have little fleas upon their backs to bite 'em,
And little fleas have lesser fleas, and so *ad infinitum*.
And great fleas themselves, in turn, have greater fleas to go on,
While these again have greater still, and greater still, and so on.

. . . while Lewis Richardson, who wrote on problems of turbulence and wind, spotted an analogy, pointing to the ultimate possibility of fleas as a generalized explanation-of-everything:

Big whorls have little whorls,
Which feed on their velocity;
And little whorls have lesser whorls,
And so on to viscosity.

Hitting the target

R. D. Clarke noted the following data for flying bombs landing on London during a certain period of the Second World War, and compared the number of hits predicted by statistical theory, with the actual number of hits. For this purpose, he divided the London area into 576 small areas each of area ¼ square kilometres. These were the results:

No. of hits	0	1	2	3	4	5 or more
Actual no. of areas hit	229	211	93	35	7	1
Predicted no. of areas hit	226.74	211.39	98.54	30.62	7.14	1.57

Euler's advice to a preacher

'Euler, the great Euler, was very pious; one of his friends, a minister of one of the Berlin churches, came to him one day and said, "Religion is lost; faith has no longer any basis; the heart is no longer moved, even by the sight of beauties, and the wonders of Creation. Can you believe it? I have represented this Creation as everything that is beautiful, poetical, and wonderful; I have quoted ancient philosophers, and the Bible itself: half the audience did not listen to me, the other half went to sleep or left the church."

"Make the experiment which truth points out to you," replied Euler. "Instead of giving the description of the world from the Greek philosophers or the Bible, take the astronomical world, unveil the world such as astronomical (*i.e.*, physical and mathematical) research constitute it. In the sermon which has been so little attended to, you have probably, according to Anaxagoras, made the sun equal to Peloponnesus. Very well! Say to your audience that, according to exact, incontestable (mathematical) measurements, our sun is 1,200,000 times larger than the earth. You have, doubtless, spoken of the fixed crystal heavens; say that they do not exist, – that comets break through them. In your explanation, planets were only distinguished from stars by movement; tell them they are worlds, – that Jupiter is 1,400 times larger than the earth, and Saturn 900 times so; describe the wonders of the ring; speak of the multiple moons of these distant worlds. Arriving at the stars, their distances, do not state miles – the numbers will be too great, they will not appreciate them; take as a scale the velocity of light; say that it travels about 186,000 miles per second; afterwards add there is no star whose light reaches us under three years, – that there are some of them with respect to which no special means of observation has been used, and whose light does not reach us under thirty years. On passing from certain results to those which have only a great probability, show that, according to all appearance, certain stars would be visible several of millions of years after having been destroyed, for the light emitted by them takes many millions of years to traverse the space which separates them from the earth."

This advice was followed; *instead of the world of fable, the minister preached the world of science*. Euler awaited the coming of his friend after the sermon with impatience. He arrived despondent, gloomy, and in a manner appearing to indicate despair. The geometer, very much astonished, cried out, "What has happened?" "Ah, Monsieur Euler," replied the minister, "I am very unhappy: they have forgotten the respect which they owed to the sacred temple, *they have applauded me*." '

Newton

When Newton saw an apple fall, he found . . .
A mode of proving that the earth turn'd round
In a most natural whirl, called gravitation;
And this is the sole mortal who could grapple
Since Adam, with a fall or with an apple.

Lord Byron

Byron was not the only versifier tempted to rhyme 'apple' with 'grapple'. The following verses appeared in *Notes & Queries* for

27 January 1887. The query, unanswered, was, who wrote them?

When Old Nick in his clutches first caught Mother Eve,
  As all the learn'd Fathers agree,
He by glozing essay'd the fair dame to deceive,
  And of knowledge he show'd her the tree.

Madam, longing to judge betwixt evil and good,
  Was curious to taste, though forbidden,
Of the fruit of life's tree, in the middle that stood,
  All erect in the Garden of Eden.

But knowledge to woman's a perilous gift,
  That unfits her too oft for her station;
Hence both Eve and poor Adam were turn'd out adrift,
  And destined to death and damnation.

Long time had this tree nearly barren remain'd,
  Unsown were its seeds in man's mind,
Till by Newton replanted it flourish'd again,
  And an apple enlighten'd mankind.

As an apple occasioned the fall of frail man,
  And with Satan compelled him to grapple,
So was knowledge decreed by the Deity's plan
  To result from the fall of an apple.

Hardy's eccentricities

'He was the classical anti-narcissist. He could not endure having his photograph taken: and there are no more than half a dozen photographs in existence. He would not have any looking-glass in his rooms, not even a shaving mirror. When he went to a hotel, his first action was to cover all the looking-glasses with towels. This would have been odd enough, if his face had been like a gargoyle: superficially it might seem odder, since all his life he was good-looking quite out of the ordinary. But, of course, narcissism and anti-narcissism have nothing to do with looks as outside observers see them.

When summer came, it was taken for granted that we should meet at the cricket ground . . . He made for his favourite place, opposite the pavilion, where he could catch each ray of sun – he was obsessively heliotropic. In order to deceive the sun into shining, he brought with him, even on a fine May afternoon, what he called his "anti-God battery". This consisted of three or four sweaters, an umbrella belonging to his sister, and a large envelope containing mathematical manuscripts, such as a Ph.D. dissertation, a paper which he was refereeing for the Royal Society, or some tripos answers. He would explain to an acquaintance that God, believing that Hardy expected the weather to change and give him a chance to work, counter-suggestibly arranged that the sky should remain cloudless.'

C. P. Snow

The End of Platonism?

'Platonism' is the belief that mathematics exists independently of the human mathematicians who study it. In other words, mathematics is discovered rather than created.

Platonism gets a strong boost from the overwhelming *feeling* that so many mathematicians experience – whatever their reason may tell them. Psychologically, Platonism makes sense. It gets a further shot in the arm from the successful use of mathematics in the sciences (see pages 87, 114). If mathematics were created by human beings, why should it turn out to be so useful in interpreting the world?

And yet . . . if mathematics does exist 'out there', then where is 'there'? Like the Heaven of Victorian Sunday School, placed above the clouds beyond the sun, its precise location is difficult to determine. Here is Eric Temple Bell, author of *The Development of Mathematics*, writing in 1940. Would he have been surprised that as his deadline approaches, mathematicians are as divided as ever?

'According to the prophets, the last adherent of the Platonic ideal in mathematics will have joined the dinosaurs by the year 2000. Divested of its mythical raiment of eternalism, mathematics will then be recognised for what it has always been, a humanly constructed language devised by human beings for definite ends prescribed by themselves. The last temple of an absolute truth will have vanished with the nothing it enshrined.'

Some letters!

MINUS QUANTITY = QUAINT TINY SUM

HIGHER MATHEMATICS = M.A. TEACHES HIM RIGHT

METRIC SYSTEM = MYSTIC METERS

MEASUREMENTS = MAN USES METER

MEASURED = MADE SURE

INNUMERABLE = A NUMBER LINE

INTEGRAL CALCULUS = CALCULATING RULES

The pleasure of proof

'My friend G. H. Hardy, who was professor of pure mathematics, . . . told me once that if he could find a proof that I was going to die in five minutes he would of course be sorry to lose me, but this sorrow would be quite outweighed by pleasure in the proof. I entirely sympathised with him and was not at all offended.'

Bertrand Russell

Kepler on the snowflake

An almost infallible sign of the highest genius is the capacity to turn creatively in many different directions. Kepler wrote 'A New Year's Gift, or On the Six-Cornered Snowflake' for his patron, 'the illustrious Counsellor at the Court of his Sacred Imperial Majesty, John Matthew Wacker'. It was published in 1611. It was the first attempt by any scientist ever to explain the genesis of such a natural object, and wonderfully displays Kepler's imaginative insight.

'For as I write it has again begun to snow, and more thickly than a moment ago. I have been busily examining the little flakes. Well, they have been falling, all of them, in radial pattern, but of two kinds: some very small with prongs inserted all the way round, indefinite in number, but of simple shapes without plumes or stripes, and very fine, but gathered at the centre into a slightly bigger globule. These formed the majority. But scattered among them were the rarer six-cornered starlets of the second kind, and not one of them was anything but flat, whether it was floating or coming to earth, with the plumes set in the same plane as their stem. Furthermore, under the flake a seventh prong inclined downwards like a root, and, as they fell, they rested on it and were held up by it for some time. This had not escaped me above, but I took it in a mistaken sense as though the three diameters were not in the same plane. So what I have said hitherto, no less than what I have had my say about, is as little removed from Nothing as may be.

The first, lumpy, kind is formed, I think, from vapour that has almost lost its heat and is on the point of condensing into watery drops. So they are round, and no beautiful shape comes their way either, abandoned as they are by the master builder [heat], and they stretch out radially in all directions on the principles that were applied above to the examination of hoar-frost formations on windows.

But in the second kind, that is, of starlets, observation of cube or octahedron has no place. There is no contact of drops, since they settle as flat objects and not, as I thought above, with three diameters crossed.

So, although here too the formative soul maintains its place and remains in play as a cause, the question of the choice of shape must be taken up again. First, why flat? Is it because I was wrong to remove, as I did, plane surfaces from among the builders of bodies? There is, after all, in all flowers a flat pentagon, not a solid dodecahedron. If so, the cause of flatness would really be this: that cold touches warm vapour on a plane and does not surround all the vapour uniformly when starlets are produced as it does when it falls in lumps.

Next, why six-cornered? Is it because this is the first of the regular figures to be essentially flat, incapable, that is, of combining with itself to form a solid body? For triangle, square, pentagon, all form bodies. Is it because the hexagon lays a flat surface without a gap? But triangle and square do the same. Or because the hexagon comes nearest to the circle of those figures which lay a flat surface without a gap? Or does this make the difference between a faculty that builds sterile shapes, triangles, and hexagons, and that second faculty that builds fruitful shapes, pentagons? Or, finally, does the nature of this formative faculty partake of six-corneredness in the inmost recess of its being?'

Ada Lovelace: how mathematicians think

Appropriately for one who was concerned to transform human calculation into mechanical, Ada Lovelace was interested in the processes by which mathematicians thought. Unfortunately, very few since have shared her interest.

'I have so many metaphysical inquiries and speculations which intrude themselves, but I am never really satisfied that I understand *anything* because, understand it as well as I may, my comprehension *can* only be an infinitesimal fraction of all I want to understand about the many connections and relations which occur to me, *how* the matter in question was first thought of and arrived at etc, etc, . . .

I am sometimes very much interested to see *how* the same conclusions are arrived at in different *ways* by different people; and I happen to be inclined to compare you and Bourdan in this case of developing exponential and logarithmic series; and very amusing it has been to me to see him *begin* exactly where you end. Your demonstration is much the best for practical purposes.'

The origins of geometry

Thales (*c.* 624–547 BC) is supposed to have visited Egypt and brought back the study of geometry, as well as discovering many propositions himself, including the proposition that 'the angle in a semi-circle is a right-angle'. This is Herodotus's account.

'The priests also said that this king divided the country among all the Egyptians, giving each an equal square plot. This was the source of his revenue, as he made them pay a fixed annual tax. If some of anyone's land was taken away by the river, he came to the king and told him what had happened. Then the king sent men to look at the land and measure how much less it was, so that in future the owner would pay the due proportion. It seems to me that land survey started from this and passed on to Greece.

The concave sundial and the division of the day into twelve were learnt by the Greeks from the Babylonians.'

'The *J*-type and the *S*-type among mathematicians'

Given the varying styles of mathematics in different countries, it was inevitable that sooner or later someone would speculate that behind them lay racial differences. Felix Klein had stated as early as 1893 that, 'a strong native space intuition seems to be an attribute of the Teutonic race, while the critical and purely logical sense is more developed in the Latin and Hebrew races'.

The following article, with the above title, was published by G. H. Hardy in *Nature*, 1934, in response to reports of Bieberbach's views. Edmund Landau, referred to in the third paragraph, was forced to resign his chair at Gottingen in 1933, following a National Socialist decree that all full-blooded Jews should lose their teaching positions.

Ironically, the Mathematical Institute at Göttingen had been financed partly by American money, $275,000 from John D. Rockefeller Jr. in 1926, and it was to America that most of the Jews who were expelled, and their non-Jewish colleagues who resigned with them, escaped.

'Mathematicians in England and America have been recently intrigued by reports of a lecture delivered by Prof. L. Bieberbach, of the University of Berlin, to the Verein zur Förderung des mathematisch-naturwissenschaftlichen Unterrichts. They have, however, found difficulty in judging the lecture fairly from second-hand reports. It is now possible to form a more reasoned estimate, Prof. Bieberbach having published a considerable extract, under the title "Persönlichkeitsstruktur und mathematisches Schaffen", in the issue of *Forschungen und Fortschritte* of June 20.

Prof. Bieberbach begins by explaining that his exposition will make clear by examples the influence of nationality, blood and race upon the creative style. For a National Socialist, the importance of this influence requires no proof. Rather is it intuitive that all our actions and thoughts are rooted in blood and race and receive their character from them. Every mathematician can recognise such influences in different mathematical styles. Blood and race determine our choice of problems, and so influence even the assured content of science (*den Bestand der Wissenschaften an gesicherten Ergebnissen*); but naturally do not go so far as to affect the value of π or the validity of Pythagoras' theorem in Euclidean geometry . . .

Our nature becomes conscious of itself in the malaise (*in dem Unbehagen*) produced by alien ways. There is an example in the manly rejection (*mannhafte Ablehnung*) of a great mathematician, Edmund Landau, by the students of Göttingen. The unGerman style of this man in teaching and research proved intolerable to German sensibilities. A people which has understood how alien lust for dominance has gnawed into its vitals . . . must reject teachers of an alien type . . .

Prof. Bieberbach proceeds to distinguish between the "*J*-type" and the "*S*-type" among mathematicians. Broadly, the *J*-type are Germans, the *S*-type Frenchmen and Jews.

Typical of the *J*-type are . . . Gauss, . . . Klein, and . . . Hilbert . . . One of the crowning achievements of the *J*-type is Hilbert's work on axiomatics, and it is particularly regrettable that abstract Jewish thinkers of the *S*-type should have succeeded in distorting it into an intellectual variety performance.

But perhaps I have quoted enough; and I feel disposed to add one comment only. It is not reasonable to criticise too closely the utterances, even of men of science, in times of intense political or national excitement. There are many of us, many Englishmen and many Germans, who said things during the War which we scarcely meant and are sorry to remember now. Anxiety for one's own position, dread of falling behind the rising torrent of folly, determination at all costs not to be outdone, may be natural if not particularly heroic excuses. Prof. Bieberbach's reputation excludes such explanations of his utterances; and I find myself driven to the more uncharitable conclusion that he really believes them true.'

The oldest puzzle in the world

'There are seven houses each containing seven cats. Each cat kills seven mice and each mouse would have eaten seven ears of spelt. Each ear of spelt would have produced seven hekats of grain. What is the total of all these?

This puzzle, freely paraphrased here, is problem 79 in the Rhind papyrus, our richest source for ancient Egyptian mathematics, which is named after the Scottish Egyptologist A. Henry Rhind, who purchased it in 1858 in Luxor.

The Rhind papyrus is in the form of a scroll about eighteen and a half feet long and thirteen inches wide, written on both sides. It dates from about 1650 BC. The scribe's name was Ahmes, and he states that he is copying a work written two centuries earlier, so the original of the Rhind papyrus was written in the same period as another famous source of Egyptian mathematics, the Moscow papyrus, dating from 1850 BC.

Returning to the cats and mice, about 2,800 years after Ahmes, Fibonacci in his *Liber Abaci* (1202) posed this puzzle:

Seven old women are travelling to Rome, and each has seven mules. On each mule there are seven sacks, in each sack there are seven loaves of bread, in each loaf there are seven knives, and each knife has seven sheaths. The question is to find the total of all of them.

The resemblance is so strong that surely Fibonacci's problem is a direct descendant, along an historical path that we can no longer trace, of the Rhind puzzle? Not necessarily. There is an undoubted fascination with geometrical series, and the number 7 is not only as magical and mysterious as any number can be, but was especially easy for the Egyptians to handle, because they multiplied by repeated doubling, and 7 = 1 + 2 + 4. Put these factors together, and you naturally arrive at two similar puzzles.

The St Ives Riddle

As I was going to St Ives,
I met a man with seven wives.
Every wife had seven sacks,
Every sack had seven cats,
Every cat had seven kits;
Kits, cats, sacks and wives,
How many were going to St Ives?

This rhyme appears in the eighteenth-century *Mother Goose* collection. Is it also descended from the Rhind papyrus and Fibonacci?'

Did you know?

There is only one long division extant in the entire corpus of Greek mathematics.

The term 'million' comes from an Italian word coined by Marco Polo.

Cardan described negative numbers as 'fictions' and their square roots as 'sophistic', and a complex root of a quadratic, which he had calculated, as being 'as subtle as it is useless'.

The symbol for infinity was used by the Romans for 1,000.

The extra energy consumed while doing difficult mental arithmetic problems would be met by eating one salted peanut every two hours.

Newton's annotated copy of Barrow's *Euclid* was sold at auction in 1920 for five shillings. Shortly thereafter it appeared in a dealer's catalogue marked at £500.

Newton is on record as speaking only once when a member of parliament, to ask that a window be opened.

George Bidder, the Calculating Boy, on himself

George Bidder (1806–1878) was unusual among calculating prodigies in using his abilities in his profession (he became a distinguished civil engineer) and in being able to explain the methods he used.

'As nearly as I recollect it was about the age of six years that I was first introduced to the science of figures . . . My first and only instructor in figures was my elder brother a working mason . . . ; the instruction he gave me was commenced by teaching me to count up to 10. Having accomplished this, he induced me to go on to 100, and there he stopped . . . at this time I did not know one written or printed figure from another, and my knowledge of language was so restricted, that I did not know there was such a word as "multiply"; but . . . I set about, in my own way, to acquire the multiplication table.

. . . there resided, in a house opposite to my father's, an aged blacksmith, a kind old man, who, not having any children, had taken a nephew as his apprentice. With this old gentleman I struck up an early acquaintance, and was allowed the privilege of running about his workshop. As my strength increased, I was raised to the dignity of being permitted to blow the bellows for him, and on winter evenings I was allowed to perch myself on his forge hearth, listening to his stories. On one of these occasions, somebody by chance mentioned a sum; . . . I gave the answer correctly. This occasioned some little astonishment; they then asked me other questions, which I answered with equal facility. They then went on to ask me up to two places of figures . . . of course I did not do it then as rapidly as afterwards, but I gave the answer correctly, as was verified by the old gentleman's nephew, who began chalking it up to see if I was right. As a natural consequence this increased my fame still more, and what was better, it eventually caused halfpence to flow into my pocket; which, I need not say, had the effect of attaching me still more to the science of arithmetic and thus by degrees I got on, until the multiple arrived at thousands.'

Problema and *theorema*

The Greeks, for whom mathematics was essentially geometry, distinguished in an instructive manner between a *problema* and a *theorema*:

'Those who favour a more exact terminology in the subjects studied in geometry . . . use the term problem to mean an inquiry in which it is proposed to do or to construct something, and the term theorem an inquiry in which the consequences and necessary implications of certain hypotheses are investigated . . .'

The distinction became a commonplace in modern Europe with the translation of Euclid into Latin. Billingsley wrote in 1570: 'A Probleme, is a proposition which requireth some action or doing . . . A Theoreme, is a proposition, which requireth the searching out and demonstration of some propertie . . . of some figure,' and Charles Hutton in 1815 repeated Billingsley almost exactly.

Godfrey and Siddons, in *The Teaching of Elementary Mathematics* (1946), however, make no mention of the distinction.

Ironically, with the introduction of computers, the distinction is once again important – many computer *problems* invite a constructive solution, and do not work less well because they involve no proof of propositions.

Turing

Turing, apart from creating one of the greatest contributions to the theory of computing, the Turing machine, was the author of a classic scientific 1952 paper in which he pointed out, in effect, that you could create structural patterns in a chemical system provided you had two reactants which diffused at different rates, if the faster component inhibited reaction while the slower component promoted it.

Only recently have chemists been able to verify his claims experimentally, creating chemical reactions in which 'spots replicate, grow and die in uncanny resemblance to living things' – or Conway's Game of Life.

Turing lived at the Crown Inn, Shenley Brook End, from 1938 to 1944. Somewhere near here he buried two silver bars, but he forgot where he buried them and they have never been recovered. The Crown is now a private house and the area where he buried the bars is a housing estate.

The natural history of differential equations

The natural development of mathematics is from the study of particular and striking more-or-less concrete objects to the general and abstract structure in which their relations to each other are clearly visible. As recently as 1958 George Temple could write of:

'that Cinderella of pure mathematics – the study of differential equations. The closely guarded secret of this subject is that it has not yet attained the status and dignity of a science, but still enjoys the freedom and freshness of such a pre-scientific study as natural history compared with botany. The student of differential equations – significantly he has no name or title to rank with the geometer or analyst – is still living at the stage when his main tasks are to collect specimens, to describe them with loving care, and to cultivate them for study under laboratory conditions. The work of classification and systematization has hardly begun. This is true even of differential equations which belong to the genus technically described as 'ordinary, linear equations' . . . In the case of non-linear equations . . . An inviting flora of rare equations and exotic problems lies before a botanical excursion into the non-linear field.'

Holy Relic

'In 1816 a tooth belonging to Sir Isaac Newton was sold in London for £730. It was purchased by a nobleman who had it set in a ring, which he wore constantly.'

Vipers, logs and all that

'It is widely believed that the only mathematician in the Bible was Noah. Nobody else would have had a hope of passing the Eleven Plus. Admittedly, Moses' Book of Numbers is frankly disappointing, but I hope to show in this article that the Bible contains evidence of a higher standard of mathematics than is generally supposed.

Arithmetic is, of course, mentioned most frequently, and we are told that men sometimes worshipped figures.[1] At a very early stage "men began to multiply,"[2] and Abraham was familiar with division.[3] Some writers have pointed out that the arithmetic in Ezra[4] is faulty, but this is explained where it reads "certain additions were made of thin work".[5] The approximation for π is reasonable,[6] considering the fact that Moses destroyed the tables,[7] which were not replaced until Solomon's time.[8] Elsewhere we read "he shall not extract the root thereof",[9] and "we wrestle against powers".[10]

The first attempts at Geometry were, of course, Euclidean. We read that "great rulers were brought down",[11] "from Syracuse they fetched a compass",[12] and Noah constructed an arc[13] and Ezekiel described a line.[14] Further progress was made when they took axes,[15] culminating in David's success with the calculus.[16] David, incidentally, was the first to refuse to accept what he had not proved.[17] St Paul was familiar with four dimensions,[18] and Joshua continued with the arc along a Jordan path.[19]

Algebra although thought to be an invention of the Arabs, was only too familiar to the Jews. For instance, Moses gives instructions about a matrix[20] and Ezekiel knew enough about rings to describe them as "dreadful".[21] Peter was kept half the night by four quaternions,[22] and the Jews were described as "a generation seeking after a sign".[23]

"As for the Pure, his work is right" said the writer of Proverbs,[24] and this attitude is reflected in the few existing references to Mathematics. "I have seen thy abominations in the Fields" cried Jeremiah,[25] and the Psalmist complained "Thou hast afflicted me with all thy Waves."[26] Later the Father of Publius was "sick of the bloody Flux".[27]

It is easy to understand why they disliked mathematics. Apart from the deacons "who purchase to themselves a good Degree",[28] they had to be examined, as was St Paul.[29] We know that Elisha passed,[30] and Solomon was able to answer all the questions,[31] but Peter was much troubled when he saw the sheet,[32] and Job cried "My kinsfolk have failed, and my friends."[33] Perhaps Jehoiakim was an examiner, for "when he had read three or four pages he cast it into the fire".[34] As for St John, all that he knew was "the Second woe is past, the Third cometh".[35]

[1]Acts vii. 43. [2]Gen. vi. 1. [3]Gen. xv. 10. [4]Ezra ii. [5]1 Kings vii. 29. [6]2 Chron. iv. 2. [7]Exod. xxxii. 19. [8]2 Chron. iv. 8. [9]Ezek, xvii. 9. [10]Eph. vi. 12. [11]Ps. 136. 17. [12]Acts xxviii. 13. [13]Gen. vi. (archaic spelling). [14]Ezek. xl. [15]1 Sam. xiii. 21. [16]1 Sam. xvii. [17]1 Sam. xvii. 39. [18]Eph. iii. 18. [19]Joshua iii. [20]Exod. xxxiv. 19. [21]Ezek. i. 18. [22]Acts xii. 4. [23]Math. xvi. 4. [24]Prov. xxi. 8. [25]Jer. xiii. 27. [26]Ps. 88. 7. [27]Acts xxviii. 8. [28]1 Tim. iii. 13. [29]Acts xxviii. 18. [30]2 Kings iv. 8. [31]2 Chron. ix. 2. [32]Acts xi. [33]Job xix. 14. [34]Jer. xxxvi. 23. [35]Rev. xi. 14.'

G. J. S. Ross

Rabbi Solomon's problem

Mathematical problems have had many sources: this is one of the most unusual.

'According to the Jewish dietary laws, foods are divided into pure and impure. The latter include also certain incompatible mixtures that are in themselves pure. The law also prescribes that the addition of an impure substance or an incompatible ingredient, provided that it does not exceed a certain fraction of the total, in most cases $\frac{1}{60}$, does not make the resultant mixture impure or incompatible. A vessel whose walls have absorbed an impurity, does not contaminate the food that is cooked in it, if it may be ascertained that the absorbed element forms no more than $\frac{1}{60}$ of the volume. Hence the impurity may be entirely disregarded if the volume of the vessel is sixty times the volume of its walls and base, for in such a case, the impurity would not exceed the prescribed limit even under the assumption that the vessel absorbed the impurity to the full extent of its walls.

This led in many cases to the discussion of the ratio of the volume of vessels to that of their walls, for it is obvious that the minimum absorption will take place when the volume of the walls is also a minimum. The reasoning implied is analogous to that used in the solution of problems in maxima and minima.

The problem treated by Rabbi Solomon consists of two parts. One shows how an open vessel with a square base may be constructed in the most economical way. The second shows how to find the dimensions of the walls if the volume of the vessel is to be sixty times the volume of the base and walls.

Rabbi Solomon expresses the result of his analysis [of the first part] in a formula which, he claims, no one before him had discovered; namely, $b = 2h$, where b is the length of a side of the base and h is the height of the vessel.'

Soroban versus Electric Calculator

The Magic Calculator, published in Japan in 1964, speculated that the *soroban*, as they call their version of the *abacus* would one day, like judo and the game of go, eventually spread outside Japan, leading to international competitions, and even a World Abacus Championship.

'When the Pacific War ended, Japan was forced to undergo tremendous changes in every field in every respect.

Japanese abacus operators got a dramatic boost in morale in

their start toward recovery when the abacus won over the electric calculator in an exciting contest.

The contest between the abacus and the electric calculator, probably planned as entertainment, was held at the Ernie Pyle Theater (Now the Takarazuka Theater, Tokyo) on November 11, 1946.

Representing the U.S. Occupation Forces in Tokyo was Pvt. Thomas N. Wood, and the Japanese representative was Mr Kiyoshi Matsuzaki of the Savings Bureau, Ministry of Postal Administration.

The two experts competed in addition, subtraction, multiplication, division and composite problems – and the abacus beat the electric computer by 4 to 1.

The problems, typewritten and mimeographed, included 50 addition and subtraction problems of three to six digits, and multiplication and division of 6 to 12 digits; and only in straight multiplication did the Japanese abacus lose out.'

Has the abacus subsequently been eclipsed by the electronic calculator? Surprisingly, not quite. In 1977 a Soroban Curriculum Institute was founded in California to encourage the use of the abacus in the classroom. It is still fastest at basic addition and subtraction – in 1981 the *soroban* champion of Taiwan, South Korea and Japan could add and subtract 15 eleven-digit numbers in 9 seconds, which is less time than it would take to key the numbers into an electronic calculator.

The *Pons Asinorum*

The fifth proposition in the first book of Euclid's *Elements* asserts that the base angles of an isosceles triangle are equal. Here is the figure that Euclid used.

In the days when every schoolboy studied Euclid, it was commonly called the *Pons Asinorum* or Ass's Bridge. The charitable explanation of this name was that the figure resembles a ramp, which an ass, but not a horse, would be able to cross. The more plausible explanation is that it refers to the unfortunate pupils for whom this was the first difficult proof in Euclid. (Euclid only used such a complicated proof, because he proved this theorem so early, before he had proved any theorems about congruent triangles.)

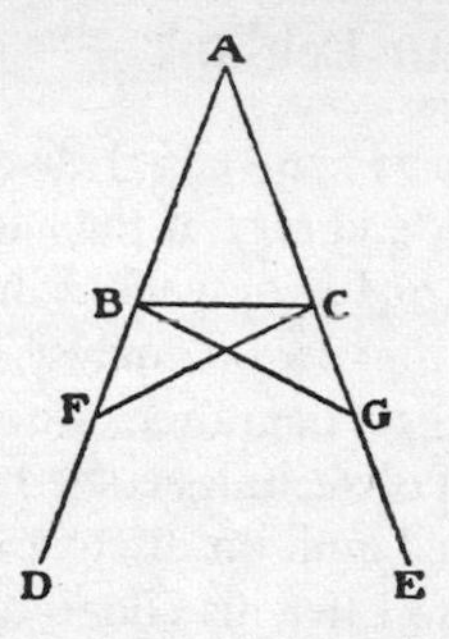

The asses among them stuck here, though the *Self-examinations in Euclid* (1829), as optimistic as its title suggests, claimed that 'men, of small intellect, will easily get over, if they will proceed "with measured steps and slow"'.

On the other hand, it could be a barb aimed at those who spent their time on geometry, while the wiser found other occupations. The term first appears in Murray's *English Dictionary* (1780) with this epigram:

> If this be rightly called the bridge of asses,
> He's not the fool that sticks but he that passes.

The view that geometry might possibly be less than fascinating appears in this modern limerick:

> In the Greek mathematical forum
> Your Euclid was present to bore 'em.
> He spent all his time
> Drawing circles sublime
> And crossing the Pons Asinorum.

The first computer program to prove geometrical theorems was described in 1959. Its most famous achievement was a very simple proof of the *Pons Asinorum* theorem. The computer added no construction lines, but argued, basically, that triangle ABC is congruent to triangle ACB, by 'two sides and the included angle'. Therefore, the angles B and C are equal. Was the computer the originator of this simple proof? It is a nice idea, but false. The same proof was known to Proclus in the fifth century.

Gauss and Monsieur Leblanc

Marie-Sophie Germain (1776–1831) discovered mathematics in the books of her father's library at the age of thirteen, and spent the period of the Reign of Terror teaching herself differential calculus. In 1795 she wrote a paper on analysis, and submitted it under the name of 'LeBlanc' to Lagrange, whose lecture notes at the newly-founded Ecole Polytechnique she had managed to obtain. Lagrange was impressed and, on discovering that the author was a woman, visited her to offer his congratulations.

In 1804 she wrote to Gauss and sent samples of her work in number theory, using the same pseudonym. Gauss replied, but without enthusiasm, until he discovered that 'Monsieur Leblanc' was a woman, whereupon he wrote:

'The taste for the abstract sciences in general and, above all, for the mysteries of numbers, is very rare: this is not surprising, since the charms of this sublime science in all their beauty reveal themselves only to those who have the courage to fathom them. But when a woman, because of her sex, our customs and prejudices, encounters infinitely more obstacles than men in familiarising herself with their knotty problems, yet overcomes these fetters and penetrates that which is most hidden, she doubtless has the most noble courage, extraordinary talent, and superior genius.'

The Taylor series remainder

'During the Russian revolution, the mathematical physicist Igor Tamm was seized by anti-communist vigilantes at a village near Odessa where he had gone to barter for food. They suspected he was an anti-Ukrainian communist agitator and dragged him off to their leader.

Asked what he did for a living, he said he was a mathematician. The sceptical gang leader began to finger the bullets and grenades slung round his neck. "All right," he said, "calculate the error when the Taylor series approximation to a function is truncated after n terms. Do this and you will go free. Fail and you will be shot." Tamm slowly calculated the answer in the dust with his quivering finger. When he had finished, the bandit cast his eye over the answer and waved him on his way.

Tamm won the 1958 Nobel prize for physics but he never did discover the identity of the unusual bandit leader.'

Free market maths

'In recent years the problem of funding in academic circles has become one of considerable size. Obviously, in maths, partnerships with industry are not always feasible. Many of the other conventional solutions are also difficult to implement. So what other possibilities are there in this area? Where can we turn to combat this result of current economic policy?

One answer which I think has been seriously overlooked is that of theorem sponsorship. For instance, think of the money that could be generated by the Coca-Cola Theory of Relativity, or the Pepsi–Bolzano–Weierstrass theorem. Organisations could be charged a small amount for each lecture or paper involving their chosen subject. This would have the advantage of generating capital for the study of those areas which are most often used. This fits very nicely with the government's current view on the free market and would no doubt be widely welcomed.

In fact, the idea has more possibilities than may be obvious at first glance. For example, companies could sponsor those subjects which are particularly relevant to their trade – "Kelvin's Economy 7 Circulation Theorem", the "Dulux White-with-a-hint-of-Green's function", or the irresistible "Remington's Fuzzaway Fuzzy Subsets".

Perhaps the most profitable area would be politics itself, surely Chaos is ideal for this. Or maybe we could have the Labour Left-half plane and the Conservative Right-half plane (though no doubt the latter would soon be privatised.)

Only one problem really springs to mind – could Sainsbury's be prosecuted by the Advertising Standards Authority for sponsoring the infinite server queue . . . ?'

Paul Norridge

Emmy Noether (1882–1935)

'In the judgement of the most competent living mathematicians, Fräulein Noether was the most significant creative mathematical genius thus far produced since the higher education of women began.'

Albert Einstein

Emmy Noether was the chief creator of modern abstract algebra, the greatest woman mathematician of all time, and one of the greatest mathematicians of the twentieth century of either sex.

Her brother, Fritz Noether, was also a mathematics professor, and her father was the distinguished mathematician Max Noether. Max once heard Emmy referred to as his daughter. Not so, he explained, 'Emmy Noether is the origin of coordinates in the Noether family.'

She had the usual problems of that period, due to her sex. Although she was the assistant and collaborator of the great Felix Klein, when Hilbert invited her to Göttingen in 1915, other members of the faculty objected. Hilbert pointed out, 'We are a university, not a bathing establishment,' but to no avail. Emmy Noether gave lectures at courses listed under Hilbert's name.

Sylvester leaves the United States in a hurry

Sylvester (1814–1897) had a sharp temper. In his youth he had been expelled from the University of London 'for taking a table knife from the refectory with the intention of sticking it into a fellow student who had incurred his displeasure'.

'In 1841 he became professor of mathematics at the University of Virginia. In almost all notices of his life nothing is said about his career there; the truth is that after the short space of four years it came to a sudden and rather tragic termination. Among his students were two brothers, fully imbued with the Southern ideas about honor. One day Sylvester criticised the recitation of the younger brother in a wealth of diction which offended the young man's sense of honor; he sent word to the professor that he must apologize or be chastised. Sylvester did not apologize, but provided himself with a sword-cane; the young man provided himself with a heavy walking-stick. The brothers lay in wait for the professor; and when he came along the younger brother demanded an apology, almost immediately knocked off Sylvester's hat, and struck him a blow on the bare head with his heavy stick. Sylvester drew his sword-cane, and pierced the young man just over the heart; who fell back into his brother's arms, calling out "I am killed." A spectator, coming up, urged Sylvester away from the spot. Without waiting to pack his books the professor left for New York, and took the earliest possible passage for England. The student was not seriously hurt; fortunately the point of the sword had struck fair against a rib.'

Chess, mathematics, and the infinite

'And what satisfaction is there comparable with a well-won "mate"? It is different from any other joy that games have to offer . . . a perfect "mate" irradiates the mind with a calm of indisputable things. It has the absoluteness of mathematics, and it gives you victory ennobled by the sense of intellectual struggle and stern justice. There are "mates" that linger in the memory like a sonnet of Keats.

It is medicine for the sick mind or the anxious spirit. We need a means of escape from the infinite, from the maze of this incalculable life, from the burden and the mystery of a world where all things "go contrary", as Mrs Gummidge used to say . . . in the midst of this infinity I know no finite world so complete and satisfying as that I enter when I take down the chessmen and marshal my knights and squires on the chequerboard field. It is then I am truly happy. I have closed the door on the infinite and inexplicable and have come into a kingdom where justice reigns, where cause and effect follow "as the night the day", and where, come victory or come defeat, the sky is always clear and the joy unsullied.'

'Alpha'

Confusion

'Johnny, what's 5 + 5?'
'It's 11, miss.'
'No, 5+5 equals 10.'
'But miss, last week 6 + 4 equalled 10!'

Paul Painlevé, politician

Paul Prudent Painlevé (1863–1933) was not the most modest of mathematicians, but, to be fair, he did not have a great deal to be modest about. Hesitating between politics, engineering and mathematicians as a career, he chose mathematics. He won the Grand Prix des Sciences Mathématiques in 1890, the Prix Bordin in 1894, and the Prix Poncelet in 1896.

In 1910 he entered politics after all. In 1917 he was Minister for War, in 1920 the Chinese government commissioned him to reorganize their country's railways, and between 1925 and 1933 he was several times minister of war and of aviation – he had earlier flown with Wilbur Wright and Henri Farman and jointly held the record for the longest duration of flight in a biplane.

'The mathematical passion is an aesthetic passion. I consider myself to be an artist. My motives have been those of any other artist with a passion for beauty. But this passion has been by no means exclusively directed towards mathematics. The mathematical passion seized upon me when I was only nine years of age. I studied with such intensity that I had reached the degree standard in mathematics by the time I was eleven. I was not a mere specialist, however, exercising an isolated faculty. I also made brilliant studies in the classics and in literature, and my passion for these subjects was quite as great as my passion for mathematics. But although all these subjects seemed to me, then, of equal importance, my teacher was not deceived. He told my father that, in spite of my interest in literature and the classics, I must nevertheless be a mathematician. But in those days the beauty of mathematics and the beauty of literature, moved me equally. I remember weeping, through excess of aesthetic delight, twice on the same day – sometime in my fifteenth year. The first occasion was produced by the description of the parting of Hector and Andromache, and the second was produced by the definition of acceleration given by Newton. Mathematics is still, to me, an aesthetic passion. I am at this moment in the middle of a short holiday. I am spending that holiday working out the mathematical theory of the gyroscope.'

A fraction of the work

'I remember as a child, in fifth grade, coming to the amazing (to me) realisation that the answer to 134 divided by 29 is $\frac{134}{29}$ (and so forth). What a tremendous labor-saving device! To me, "134 divided by 29" meant a certain tedious chore, while $\frac{134}{29}$ was an object with no implicit work. I went excitedly to my father to explain my major discovery. He told me that of course this is so, a/b and a divided by b are just synonyms. To him it was a small variation in notation.'

William Thurston, Fields Medallist 1990

The power of Calculation

'Whatever there is at all in the three worlds, which are possessed of moving and non-moving beings, cannot exist apart from Ganita (calculation).'

Mahavira (AD 850)

Fermat's Last Theorem

The press release reproduced below is self-explanatory. Unfortunately, Professor Wiles's original proof was flawed. The gap detected in the proof was, however, filled by two papers in *Annals of Mathematics*, vol. 142, 1995, the first of which alone is more than one hundred pages long and refers to the work of scores of other mathematicians – a sign of the depth and profundity of Wiles's work.

Wiles only approached the subject of Fermat's Last Theorem towards the end of the lecture referred to in the press release, as if hesitating at his own boldness. He was following the example of Yoichi Miyaoka of the Tokyo Metropolitan University, who claimed to have solved it in 1988: Miyaoka also originally presented his arguments in a lecture, without actually mentioning Fermat's Last Theorem by name at all. His proof was fatally flawed.

FERMAT'S LAST THEOREM PROVED

Mathematical result of the century

A proof of one of the most famous mathematical theorems, which scientists have been seeking for centuries, was announced in Cambridge today. The proof of Fermat's Last Theorem has been sought by mathematicians for centuries, and its solution must be considered as one of the most important mathematical results of this century.

Professor Andrew Wiles FRS of Princeton University made the announcement in a lecture at Cambridge's Isaac Newton Institute for Mathematical Sciences. Professor Wiles announced a proof of the celebrated conjecture known as Fermat's Last Theorem, which asserts that:

If n is any integer bigger than 2, the equation

$$x^n + y^n = z^n$$

has no solution for which x, y and z are all (non-zero) integers.

Fermat's Last Theorem has been proved using computers for all n less than 30,000 but until today no general proof had been found.

Professor Wiles made this morning's announcement to an audience of eminent mathematicians at the Isaac Newton Institute, which brings together mathematicians from all over the world to brainstorm about mathematical problems. Professor Wiles is currently doing research at the Institute into number theory.

Professor Wiles' work uses ideas of many mathematicians, but one of the crucial new ingredients is the work of Dr Matthias Flach of Heidelberg University. Both Professor Wiles and Dr Flach did their doctoral studies in The Department of Pure Mathematics and Mathematical Statistics at Cambridge University.

In the same lecture this morning, Professor Wiles outlined a proof of a major part of another deep conjecture in number theory (stating that every elliptic curve with rational coefficients can be parametrized by modular forms).

[Ends]

Mathematicians can be strongly persuaded by heuristic arguments, probabilistic proofs, and mere hunches. Stan Wagon wrote an article in 1986 titled 'The Evidence, Fermat's Last Theorem'. He quoted the sound result that 'Fermat's Last Theorem is true for almost all exponents', but concluded by quoting the view of Harold Edwards that 'there seems to be no reason at all to assume that Fermat's Last Theorem is true'.

In 1991 a book was published, titled *The Cabinet of Curiosities* which described, among other weird and wonderful objects, Tycho Brahe's nose, George Washington's false teeth, Jeremy Bentham's preserved corpse, which can still be seen in University College London – and Fermat's Last Theorem.

In 1908 a German mathematician called Wolfskehl achieved a certain sort of fame by leaving 100,000 marks in his will to the Academy of Science at Göttingen for the first sound proof of Fermat's Last Theorem. In those days, 100,000 marks was worth just under £5,000; its value vanished in the great German inflation of the 1930s, but for a while his well-publicized generosity produced a deluge of false arguments.

Edmund Landau had a standard form printed by means of which he replied to correspondents who claimed to have proved Fermat's Last Theorem. It read, 'On page —, lines — to — you will find a mistake.' Each form was filled in by one of his graduate students.

Marilyn vos Savant has the record-breaking IQ of 230 (and the name is genuine). Less than five months after Wiles's lecture, she published a book, *The World's Most Famous Math Problem (The Proof of Fermat's Last Theorem and Other Mathematical Mysteries)*, in which she claimed that non-Euclidean geometry is unsound, and so therefore is Wiles's proof, because he uses non-Euclidean geometry. She also claimed that his proof depends on developments in mathematics that are relatively recent and poorly understood, and she encouraged her readers to try to 'demolish Einstein's theories of relativity' by proving the parallel postulate.

The remarkable Nicolas Bourbaki

Nicolas Bourbaki is one of the hardest-working and most productive mathematicians of the twentieth century. He – he has always been entirely male – was born round about the mid-1930s, and was apparently still alive and well in the mid-1980s, rumours of his death notwithstanding.

Some of his publications have suggested that he is 'professor of the University of Nancago', but this is misleading. An article published in the American Mathematical Monthly in 1950 claimed in a footnote that professor Bourbaki was formerly of the Royal Poldavian Academy, but this is also thought to be false.

An invitation to address a Warwick University Symposium on Harmonic Analysis, in 1968, brought forth the excuse that Monsieur Bourbaki's 'well-known timidity and modesty prevented him from speaking in public'. This is also believed to be untrue.

On the other side, Claude Chevalley, one of the seven founding members, described how one of André Weil's students needed a particular result for his thesis. Weil was convinced that the required result was true, and could therefore safely be used, but he was too lazy to write out a proof. So his student quoted the result, and referred in a note to 'Nicolas Bourbaki of the Royal Academy of Poldavia'.

In Moscow in 1966, the distinguished French mathematician Jean Dieudonné stated in response to a question that, 'I respect Monsieur Bourbaki very much, but I am not, unfortunately, acquainted with him.' This is difficult to believe, since N. Bourbaki has asked the Moscow publishing house which has published Russian translations of his work, to make payments to 'my friend Jean Dieudonné'.

In the late 1940s, a paragraph about Bourbaki appeared in the Book of the Year of the *Encyclopaedia Britannica* describing him as a group, and written by one Ralph P. Boas. The *Britannica* at once received a signed letter from N. Bourbaki, protesting against this extraordinary claim. Nicolas Bourbaki then started a rumour that Ralph P. Boas did not exist, but was in fact the pseudonym of a group of young American mathematicians.

Now we are getting a little closer to the truth, or to a rough approximation to it. The name 'Bourbaki' is commonly supposed to be that of General Charles Denis Sauter Bourbaki, a nineteenth-century French general, who once turned down the offer of the throne of Greece, later tried to shoot himself while imprisoned in Switzerland, but missed and lived to the age of 83.

The 'Nancago' claim was due to the fact that at the time Dieudonné was at the University of Nancy, while his co-founder, André Weil, was by then at the University of Chicago.

It consists of between ten and twenty members, who retire from the group at age fifty. They were all initially French, including Jean Dieudonné. Michael Atiyah is a British member.

Since 1939 they have been writing a very detailed account, under the general title, *Eléments de Mathématiques*, of those parts of the mathematical landscape which they consider to be so well explored and developed that they can be presented starting from the most general ideas, in a sort of modern version of Euclid's *Elements*. So far they have published more than two dozen volumes.

A detective story

'Best of all, I read Pontrjagin's "Topological Groups". The English translation by Mrs Lehmer, (usually referred to as Emma Lemma) had just come out, and it was an eye-opener, a revelation, a thriller. Yes, a thriller – I read it almost as I would read a detective story, to find out who dunit.'

Paul Halmos

The Fool and nothing

'Now thou art an o without a figure. I am better than thou art now. I am a fool, thou art nothing.'

The Fool in *King Lear*

Hilbert on Mathematical Problems

In a lecture delivered before the International Congress of Mathematicians at Paris in 1900, David Hilbert took as his subject 'Mathematical Problems'. This passage is from his introduction. Later in the lecture he actually discussed 23 problems, some rather specific, some broader and more general, which he believed were important for mathematicians in the twentieth century. One of these is mentioned on p. 30.

'Who of us would not be glad to lift the veil behind which the future lies hidden; to cast a glance at the next advances of our science and at the secrets of its development during future centuries? What particular goals will there be toward which the leading mathematical spirits of coming generations will strive? What new methods and new facts in the wide and rich field of mathematical thought will the new centuries disclose?

History teaches the continuity of the development of science. We know that every age has its own problems, which the following age either solves or casts aside as profitless and replaces by new ones. If we would obtain an idea of the probable development of mathematical knowledge in the immediate future, we must let the unsettled questions pass before our minds and look over the problems which the science of to-day sets and whose solution we expect from the future. To such a review of problems the present day, lying at the meeting of the centuries, seems to me well adapted. For the close of a great epoch not only invites us to look back into the past but also directs our thoughts to the unknown future.

The deep significance of certain problems for the advance of mathematical science in general and the important role which they play in the work of the individual investigator are not to be denied. As long as a branch of science offers an abundance of problems, so long is it alive; a lack of problems foreshadows extinction or the cessation of independent development. Just as every human undertaking pursues certain objects, so also mathematical research requires its problems. It is by the solution of problems that the investigator tests the temper of his steel; he finds new methods and new outlooks, and gains a wider and freer horizon.

It is difficult and often impossible to judge the value of a problem correctly in advance; for the final award depends upon the gain which science obtains from the problem. Nevertheless we can ask whether there are general criteria which mark a good mathematical problem. An old French mathematician said: "A mathematical theory is not to be considered complete until you have made it so clear that you can explain it to the first man whom you meet on the street." This clearness and ease of comprehension, here insisted on for a mathematical theory, I should still more demand for a mathematical problem if it is to be perfect; for what is clear and easily comprehended attracts, the complicated repels us.

Moreover a mathematical problem should be difficult in order to entice us, yet not completely inaccessible, lest it mock at our efforts. It should be to us a guide post on the mazy paths to hidden truths, and ultimately a reminder of our pleasure in the successful solution.

The mathematicians of past centuries were accustomed to devote themselves to the solution of difficult particular problems with passionate zeal. They knew the value of difficult problems . . .

Fermat had asserted, as is well known, that the diophantine equation

$$x^n + y^n = z^n$$

(x, y and z integers) is unsolvable – except in certain self-evident cases. The attempt to prove this impossibility offers a striking example of the inspiring effect which such a very special and apparently unimportant problem may have upon science. For Kummer, incited by Fermat's problem, was led to the introduction of ideal numbers and to the discovery of the law of the unique decomposition of the numbers of a circular field into ideal prime factors – a law which to-day, in its generalization to any algebraic field by Dedekind and Kronecker, stands at the center of the modern theory of numbers and whose significance extends far beyond the boundaries of number theory into the realm of algebra and the theory of functions.'

Newton on chaos and chance

'. . . all material Things seem to have been composed of the hard and solid Particles above-mention'd, variously associated in the first Creation by the Counsel of an intelligent Agent. For it became him who created them to set them in order. And if he did so, it's unphilosophical to seek for any other Origin of the World, or to pretend that it might arise out of a Chaos by the mere Laws of Nature . . . For while Comets move in very excentrick Orbs in all manner of Positions, blind Fate could never make all the Planets move one and the same way in Orbs concentrick, some inconsiderable Irregularities excepted . . . Such a wonderful Uniformity in the Planetary System must be allowed the Effect of Choice.'

Isaac Newton

Samuel Butler at Shrewsbury School

'I have just been interrupted in this letter to prove to the whole private room, except More, that when $d = 0$, $ad = 0$, none of them being for some time able to comprehend the simple fact that four times nought is 0, and all insisting that 0 times 4 is a very different thing from 4 times 0!!!! and not above half believing that 5 times 6 = 6 times 5!!!! So much for the mathematical education at Shrewsbury: it really has been a most animated argument.'

Samuel Buttler, aged 16, writing to his mother in 1851

The road to mathematics

'By keenly confronting the enigmas that surround us, and by considering and analysing the observations I had made, I ended up in the realm of mathematics. Although I am absolutely innocent of training or knowledge in the exact sciences, I often seem to have more in common with mathematicians than with my fellow artists.'

M. C. Escher

Wiener's and von Neumann's methods of working

'Norbert Wiener used to go into a depression, and to goad himself by this depression, to goad himself emotionally into a higher level or deeper level of probing and examining and a kind of desperation . . . He would show [his depression], for he had a very expressive face and expressive voice, he would be mournful and lugubrious. He would say things like "It isn't going at all," "I can't find it." Sometimes he said, "We are wasting our time," but only if he said it cheerfully, did it mean he was giving it up. When he said it with a strong affect of misery and hopelessness, he was in the midst of it, and suddenly would become enthusiastic – "Aha, I have it" –

and write it on the board – he couldn't stand to be depressed too long. Then he might discover that his idea didn't work after all, so back into the depression. And then one of the times his "I have it, I've got it!" would really work out. And that would be it.'

According to Julian Bigelow:

'Wiener thought as a physicist really. He thought in terms of process. He thought in terms of some kinds of physical models or some kinds of intuitive models. Von Neumann was a fantastic craftsman of theory. He was also immensely imaginative . . . The craftsmanship was so strong that you weren't even aware of it, but he also had the kind of imagination that could see through the problem . . . He could write a problem out the first time he heard it, with a very good notation to express the problem . . . He was very careful that what he said and what he proved and what he wrote down was really exactly what he meant . . . Wiener didn't give a damn. For Wiener, mathematical notation and language was an encumbrance. He could get an insight and he would be wanting to say it and he would fumble with the notation . . . because he was not thinking about the notation he was working in, but he was thinking about the problem behind it . . . Von Neumann would get a new problem and analyze it all the way through to the end, and end up with a computation. He would write down the numerical computation and the order of magnitude of how big it would be, you see, and if it then didn't come out correctly, he would go back and check his steps and see where he made a mistake. Now Wiener never did that. Wiener would erase the blackboard and start and do it again, a different way, in a new notation, which didn't fit either, but he was trying to say something, which after he got all through was correct.'

Seki Kowa (1642–1708)

Seki, born in the same year as Newton, was the first and greatest master of traditional Japanese mathematics.

He introduced determinants in 1683, and gave a rule for expressing them diagrammatically, ten years before they were first considered in Europe by Leibniz.

He discovered a procedure for the approximate solution of

equations, used long before in China, that was substantially the same as Horner's (see page 142).

Although he had no notion of the derivative, he derived from an algebraic expression $f(x)$ another expression that was the equivalent to its derivative. He could also find the double roots of equations.

He calculated $\pi = 3.14159265359$, which is accurate except for the last figure.

He also discovered what we call Newton's Method for finding the approximate roots of an equation.

In one of his investigations he introduced what we call the Bernoulli numbers, which were only introduced in the West by Jakob Bernoulli in his *Ars conjectandi*, published in 1713, after Seki's death.

Vitruvian proportion

The classical world, Greek and Roman, was devoted to ideas of proportion and harmony, whether used to explain the beauty of art or the construction of the universe. Naturally, the human body partook of both these qualities, as Vitruvius explained:

'. . . in the human body the central point is naturally the navel. For if a man be placed flat on his back, with his hands and feet extended, and a pair of compasses is centred at his navel, the fingers and toes of his two hands and feet will touch the circumference of a circle described therefrom. And just as the human body yields a circular outline, so too a square figure may be found from it. For if we measure the distance from the soles of the feet to the top of the head, and then apply the measure to the outstretched arms, the breadth will be found to be the same as the height, as in the case of plane surfaces which are perfectly square.'

This passage naturally led artists to draw illustrations of 'Vitruvian man', as the figure came to be called. This example is by Leonardo.

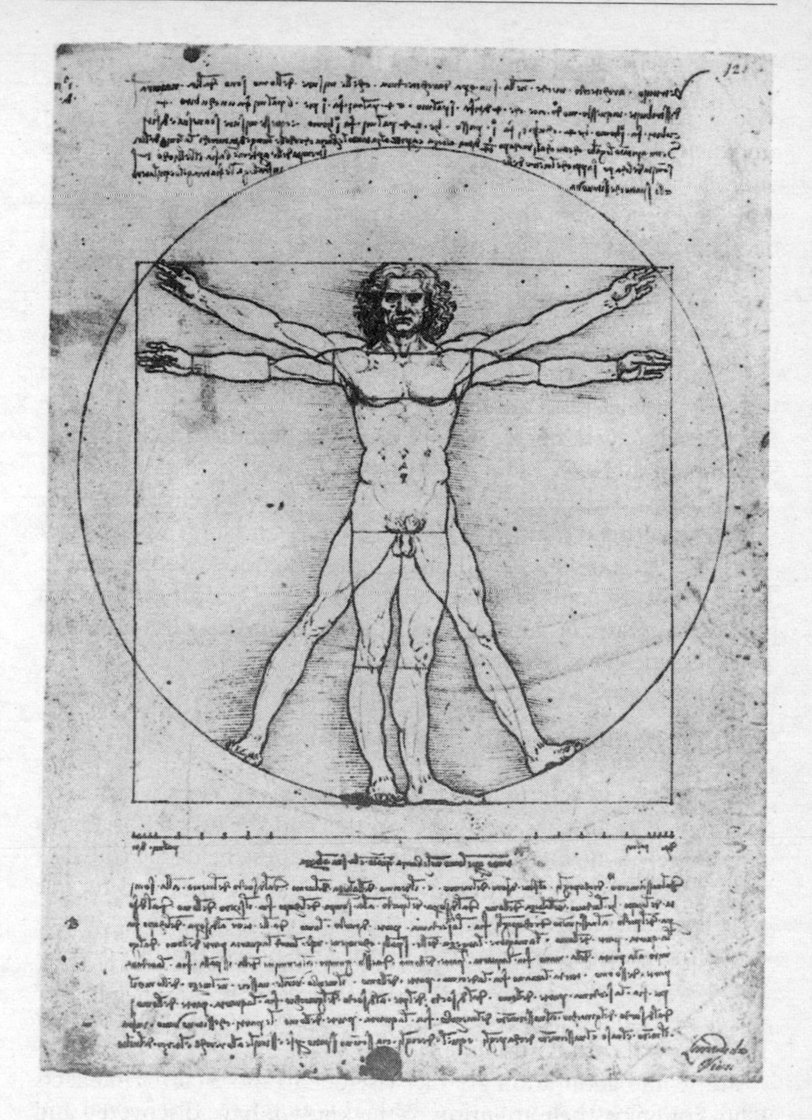

Sylvester finds a home from home

'He was now 78 years of age, and suffered from partial loss of sight and memory. He was subject to melancholy, and his condition was indeed "forlorn and desolate". His nearest relatives were nieces, but he did not wish to ask their assistance. One day, meeting a mathematical friend who had a home in London, he complained of the fare at the Club, and asked his friend to help him find suitable private apartments where he could have better cooking. They drove about from place to place for a whole afternoon, but none suited Sylvester. It grew late: Sylvester said, "You have a pleasant home: take me there," and this was done. Arrived, he appointed one daughter his reader and another daughter his amanuensis. "Now," said he, "I feel comfortably installed; don't let my relatives know where I am." The fire of his temper had not dimmed with age, and it required all the Christian fortitude of the ladies to stand his exactions. Eventually, notice had to be sent to his nieces to come and take charge of him. He died on the 15th of March, 1897, in the 83d year of his age, and was buried in the Jewish cemetery at Dalston.'

Galileo recommended by his publisher

'Since society is held together by the mutual services which men render one to another, and since to this end the arts and sciences have largely contributed, investigations in these fields have always been held in great esteem and have been highly regarded by our wise forefathers. The larger the utility and excellence of the inventions, the greater has been the honor and praise bestowed upon the inventors. Indeed, men have even deified them and have united in the attempt to perpetuate the memory of their benefactors by the bestowal of this supreme honor.

Praise and admiration are likewise due to those clever intellects who, confining their attention to the known, have discovered and corrected fallacies and errors in many and many a proposition enunciated by men of distinction and accepted for ages as fact. Although these men have only pointed out falsehood and have not replaced it by truth, they are nevertheless worthy of commendation when we consider the well-known difficulty of discovering fact, a difficulty which led the prince of orators to exclaim: *Utinam tam facile possem vera reperire, quam falsa convincere*. And indeed, these latest centuries merit this praise because it is during them that the arts and sciences discovered by the ancients, have been reduced to so great and constantly increasing perfection through the investigations and experiments of clear-seeing minds. This development is particularly evident in the case of the mathematical sciences. Here, without mentioning various men who have achieved success, we must without hesitation and with the unanimous approval of scholars assign the first place to Galileo Galilei, Member of the Academy of the Lincei. This he deserves not only because he has effectively demonstrated fallacies in many of our current conclusions, as is amply shown by his published works, but also because by means of the telescope (invented in this country but greatly perfected by him) he has discovered the four satellites of Jupiter, has shown us the true character of the Milky Way, and has made us acquainted with spots on the Sun, with the rough and cloudy portions of the lunar surface, with the threefold nature of Saturn, with the phases of Venus and with the physical character of comets. These matters were entirely unknown to the ancient astronomers and philosophers; so that we may truly say that he has restored to the world the science of astronomy and has presented it in a new light.

Poisson and the pendulum

Siméon-Denis Poisson (1781–1840) specialized in applications of mathematics, and teaching. He is supposed to have once said that there are only two good things in life, doing mathematics and teaching it.

'The boy was put out to nurse, and he used to tell how one day his father, coming to see him, found that the nurse had gone out, on pleasure bent, having left him suspended by a small cord attached to a pail fixed in the wall. This, she explained, was a necessary precaution to prevent him from perishing under the teeth of the various animals and animalculae that roamed on the floor. Poisson used to add that his gymnastic efforts carried him incessantly from one side to the other, and it was thus in his tenderest infancy that he commenced those studies on the pendulum that were to occupy him so large a part of his mature age.'

W. W. Rouse Ball

The Final Oral Exam

A duet between the Professor, P, and the Student, S.

'P: There is grandeur in the grading of a group,
There is beauty in a singularity;
There's an elegance emergent
In an integral divergent
And the counting of the edges of a tree.

S: Yes, I love to practice counting
To enormous sums amounting
And especially on the edges of a tree.

P: Categories have a splendor that is grim,
And functors always soothe the troubled mind;
And to him that's mathematic
There is nothing too erratic
In a process that's stochastically defined.

S: Yes, despite its oscillation
I can find the expectation
Of a process that's stochastically defined.

P&S: If that is so, sing derry down derry,
It's evident, very, that $\frac{\text{you are}}{\text{I am}}$ done.
Away $\frac{\text{you'll}}{\text{I'll}}$ go with theorem and query
And corollary, till tenure's won.

P: Can you rectify a sinusoidal arc?
Do you fancy you are erudite enough?
Information I'm requesting
On the subject I am testing:
Can you tell me more about the classic stuff?

S: The professors that I see, Sir,
Universally agree, Sir,
That you never have to know the classic stuff.

P: Do you have enough results to graduate?

Will you try to prove a theorem on your own?
The market is terrific
For an author that's prolific:
Do you think you are sufficiently well known?

S: To the matter that you mention
I have given some attention,
And I think I am sufficiently well known.

P&S: If that is so, sing derry down derry,
It's evident, very, that $\frac{\text{you are}}{\text{I am}}$ done.
Away $\frac{\text{you'll}}{\text{I'll}}$ l go with theorem and query
And corollary, till tenure's won.'

C. L. Fefferman and G. B. Folland

Euler as a marine engineer

Archimedes, the greatest mathematician of antiquity, constructed engines of war to defend Syracuse against the Romans. Ever since, mathematicians have been used by governments, whether willingly or unwillingly, to improve the 'arts of war'.

'On August 23, 1774, within a month of his appointment as Ministre de la Marine and the day before he was made Comptroleur Général of France, Turgot wrote as follows to Louis XVI:

The famous Leonard Euler, one of the greatest mathematicians of Europe, has written two works which could be very useful to the schools of the Navy and the Artillery. One is a *Treatise on the Construction and Manoeuvering of Vessels*; the other is a commentary on the principles of artillery of Robins . . . I propose that Your Majesty order these to be printed; . . .

It is to be noted that an edition made thus without the consent of the author injures somewhat the kind of ownership he has of his work. But it is easy to recompense him in a manner very flattering for him and glorious to Your Majesty. The means would be that Your Majesty would vouchsafe to authorize me to write on Your Majesty's part to the lord Euler and to cause him to receive a gratification equivalent to what he could gain from the edition of his book, which would be about 5,000 francs. This sum will be paid from the secret accounts of the Navy.'

The discovery of quaternions

'If I may be allowed to speak of *myself* in connexion with the subject, I might do so in a way which would bring *you* in, by referring to an *ante-quaternionic* time, when you were a mere *child*, but had caught from me the conception of a Vector, as represented by a *Triplet*: and indeed I happen to be able to put the finger of memory upon the year and month – October, 1843 – when having recently returned from visits to Cork and Parsonstown, connected with a Meeting of the British Association, the desire to discover the laws of the multiplication referred to regained with me a certain strength and earnestness, which had for years been dormant, but was then on the point of being gratified, and was occasionally talked of with you. Every morning in the early part of the above-cited month, on my coming down to breakfast, your (then) little brother William Edwin, and yourself, used to ask me, "Well, Papa, can you *multiply* triplets?" Whereto I was always obliged to reply, with a sad shake of the head: "No, I can only *add* and subtract them."

But on the 16th day of the same month – which happened to be a Monday, and a Council day of the Royal Irish Academy – I was walking in to attend and preside, and your mother was walking with me, along the Royal Canal, to which she had perhaps driven; and although she talked with me now and then, yet an *under-current* of thought was going on in my mind, which gave at last a *result*, whereof it is not too much to say that I felt *at once* the importance. An *electric* current seemed to *close*; and a spark flashed forth, the herald (as I *foresaw, immediately*) of many long years to come to definitely directed thought and work, by *myself* if spared, and at all events on the part of *others*, if I should even be allowed to live long enough distinctly to communicate the discovery. Nor could I resist the impulse – unphilosophical as it may have been – to cut with a knife on a stone of Brougham Bridge, as we passed it, the fundamental formula with the symbols, i, j, k; namely

$$i^2 = j^2 = k^2 = ijk = -1$$

Hamilton

The ocean of truth

'I do not know what I may appear to the world, but to myself I seem to have been only like a boy playing on the sea-shore, and diverting myself in now and then finding a smoother pebble or a prettier shell than ordinary, while the great ocean of truth lay all undiscovered before me.'

Isaac Newton, shortly before his death

'With the help of the accomplished holy sages, who are worthy to be worshipped by the lords of the world . . . I glean from the great ocean of the knowledge of numbers a little of its essence, in the manner in which gems are [picked] from the sea, gold from the stony rock and the pearl from the oyster shell; and I give out according to the power of my intelligence, the *Sara Samgraha*, a small work on arithmetic, which is [however] not small in importance.'

Mahavira (*c.* 850)

Erdös lends a hand

'I knew very little number theory at the time, and I tried to find a proof along purely probabilistic lines but to no avail. In March 1939 I journeyed from Baltimore to Princeton to give a talk. Erdös, who was spending the year at the Institute for Advanced Study, was in the audience but he half-dozed through most of my lecture; the subject matter was too far removed from his interests. Toward the end I described briefly my difficulties with the number of prime divisors. At the mention of number theory Erdös perked up and asked me to explain once again what the difficulty was. Within the next few minutes, even before the lecture was over, he interrupted to announce that he had the solution!

The final result for the number of prime divisors $v(m)$ is as follows: The proportion (density) of integers m for which

$$\log \log m + a\sqrt{2 \log \log m} < v(m) < \log \log m + b\sqrt{2 \log \log m}$$

is given by the area under the normal curve

$$\frac{1}{\sqrt{\pi}} e^{-x^2}$$

between $x = a$ and $x = b$.

The reader, I hope, will forgive my lack of modesty if I say that it is a beautiful theorem. It marked the entry of the normal law, hitherto the property of gamblers, statisticians and *observateurs*, into number theory and, as I said earlier, it gave birth to a new branch of this ancient discipline.

It took what looks now like a miraculous confluence of circumstances to produce our result. Each of us contributed something which was almost routine in our respective areas of competence and neither of us was familiar with the ingredients which the other had in his possession and which were all essential for success.

It would not have been enough, certainly not in 1939, to bring a number theorist and a probabilist together. It had to be Erdös and me: Erdös because he was almost unique in his knowledge and understanding of the number theoretic method of Viggo Brun, which was the decisive and, I may add, the deepest of the ingredients, and me because I could see independence and the normal law through the eyes of Steinhaus.'

Mark Kac

Mathematicians, the users and makers of signs

'One of the secrets of analysis consists in the characteristic, that is, in the art of skillful employment of the available signs.'

Leibniz

'Chen Ning Yang, the Nobel prize physicist, tells a story which illustrates an aspect of the intellectual relation between mathematicians and physicists at present:

One evening a group of men came to a town. They needed to have their laundry done so they walked around the city streets trying to find a laundry. They found a place with the sign in the window, "Laundry Taken in Here". One of them asked: "May we leave our laundry with you?" The proprietor said: "No. We don't do laundry here." "How come?" the visitor asked. "There is such a sign in your window." "Here we make signs," was the reply. This is somewhat the case with mathematicians. They are the makers of signs which they hope will fit all contingencies. Yet physicists have created a lot of mathematics.

In some of the more concrete parts of mathematics – for example probability theory – physicists like Einstein and Smoluchowski have opened certain new areas even before mathematicians. The ideas of information theory, of entropy of information and its role in general continuum originated with physicists like Leo Szilard and an engineer, Claude Shannon, and not with "pure" mathematicians who could and ought to have done so long before.'

Stanislav Ulam

Pythagoras before Pythagoras

A tablet from Susa, dating from the period 1900–1650 BC, uses Pythagoras' Theorem to find the circumradius of a triangle whose sides are 50, 50 and 60. Pythagoras himself lived in the sixth century BC.

Hilbert as a mathematical physicist

'Already before Minkowski's death in 1909, Hilbert had begun a systematic study of theoretical physics, in close collaboration with his friend, who had [also] always kept in touch with the neighboring science. Minkowski's work on relativity theory was the first fruit of these joint studies. Hilbert continued them through the years, and between 1910 and 1930 often lectured and conducted seminars on topics of physics. He greatly enjoyed this widening of his horizon and his contact with physicists, whom he could meet on their own ground. The harvest however can hardly be compared with his achievements in pure mathematics. The maze of experimental facts which the physicist has to take into account is too manifold, their expansion too fast, and their aspect and relative weight too changeable for the axiomatic method to find a firm enough foothold, except in the thoroughly consolidated parts of our physical knowledge. Men like Einstein or Niels Bohr grope their way in the dark toward their conceptions of general relativity or atomic structure by another type of experience and imagination than those of the mathematician, although no doubt mathematics is an essential ingredient. Thus Hilbert's vast plans in physics never matured.'

A South American abacus

'In order to carry out a very difficult computation for which an able computer would require paper and pen, these Indians make use of their kernels. They place one here, three somewhere else and eight, I know not where. They move one kernel here and there and the fact is that they are able to complete their computation without making the smallest mistake. As a matter of fact, they are better at practical arithmetic than we are with pen and ink. Whether this is not ingenious and whether these people are wild animals, let those judge who will! What I consider as certain is that in what they undertake to do they are superior to us.'

José de Acosta, on the Incas

Gödel and Kafka

'Kurt Gödel was unquestionably the greatest logician of the century. He may also have been one of our greatest philosophers. When he died in 1978, one of the speakers at his memorial service made a provocative comparison of Gödel with Einstein . . . and with Kafka.

Like Einstein, Gödel was German-speaking and sought a haven from the events of the Second World War in Princeton. And like Einstein, Gödel developed a structure of exact thought that forces everyone, scientist and layman alike, to look at the world in a new way.

The Kafkaesque aspect of Gödel's work and character is expressed in his famous Incompleteness Theorem of 1930. Although this theorem can be stated and proved in a rigorously mathematical way, what it seems to say is that *rational thought can never penetrate to the final, ultimate truth*. A bit more precisely, the Incompleteness Theorem shows that human beings can never formulate a correct and complete description of the set of natural numbers, {0, 1, 2, 3 . . . }. But if mathematicians cannot ever fully understand something as simple as number theory, then it is certainly too much to expect that science will ever expose any ultimate secret of the universe.

Scientists are thus left in a position somewhat like K. in *The Castle*. Endlessly we hurry up and down corridors, meeting people, knocking on doors, conducting our investigations. But the ultimate success will never be ours. Nowhere in the castle of science is there a final exit to absolute truth.

This seems terribly depressing. But, paradoxically, to understand Gödel's proof is to find a sort of liberation. For many logic students, the final breakthrough to full understanding of the Incompleteness Theorem is practically a conversion experience. This is partly a by-product of the potent mystique Gödel's name carries. But, more profoundly, to understand the essentially labyrinthine nature of *the castle* is, somehow, to be free of it.'

Rudy Rucker

Sylvester's enthusiasm

'The one thing which constantly marked Sylvester's lectures was enthusiastic love of the thing he was doing. He had in the fullest possible degree, to use the French phrase, the defect of this quality; for as he almost always spoke with enthusiastic ardor, so it was almost never possible for him to speak on matters incapable of evoking this ardor. In other words, the substance of his lectures had to consist largely of his own work, and, as a rule, of work hot from the forge. The consequence was that a continuous and systematic presentation of any extensive body of doctrine already completed was not to be expected from him. Any unsolved difficulty, any suggested extension, such as would have been passed by with a mention by other lecturers, became inevitably with him the occasion of a digression which was sure to consume many weeks, if indeed it did not take him away from the original object permanently. Nearly all of the important memoirs which he published, while in Baltimore, arose in this way. We who attended his lectures may be said to have seen these memoirs in the making. He would give us on the Friday the outcome of his grapplings with the enemy since the Tuesday lecture. Rarely can it have fallen to the lot of any class to follow so completely the workings of the mind of the master.'

The numerical record

'The earliest evidence of a numerical recording device is a section of a fibula of a baboon, with 29 clearly visible notches, dated to about 35,000 BC, from a cave in the Lebembo Mountains on the borders of Swaziland in southern Africa.'

A non-obvious conclusion

'Rules for Demonstration. 1. Do not attempt to demonstrate any of those things so self-evident that we have nothing clearer to prove them by.'

Pascal

'I am afraid that what seems to you simple is to [the average boy] complex, and what seems to you complex is to him quite simple.'

John Perry (1909)

Nature, art and microscopes

Why are fractals so popular? No doubt a contributory cause is their *natural* appearance, swirling, twisting, turning, Romantic in the extreme. In fractal images mathematicians create pictures only seen previously in nature. John Wilkins was a founding member of the Royal Society, projector of human beings flying to the moon, and Bishop of Chester (1614–1672).

'I cannot here omit the Observations which have been made in these later times since we have had the use and improvement of the Microscope, concerning that great difference which by the help of that doth appear, betwixt *natural* and *artificial* things. Whatever is Natural doth by that appear, adorned with all imaginable Elegance and Beauty. There are such inimitable Gildings and Embroideries in the smallest Seeds of Plants, but especially in the parts of Animals, in the Head or Eye of a small Fly; Such accurate Order and Symmetry in the frame of the most minute Creatures, a Lowse or a Mite, as no man were able to conceive without seeing of them. Whereas the most curious Works of Art, the sharpest and finest Needle doth appear a blunt rough Bar of Iron, coming from the Furnace of the Forge. The most accurate engravings or embossments seem such rude bungling deformed Works, as if they had been done with a Mattock or a Trowel. So vast a difference is there between the Skill of Nature, and the rudeness and imperfection of Art.'

John Wilkins

Arnol'd on reading mathematics

'It is almost impossible for me to read contemporary mathematicians who, instead of saying, "Petya washed his hands," write simply: "There is a $t_1 < 0$ such that the image of t_1 under the natural mapping $t_1 \rightarrow \text{Petya}(t_1)$ belongs to the set of dirty hands, and a t_2, $t_1 < t_2 \leq 0$, such that the image of t_2 under the above-mentioned mappings belongs to the complement of the set defined in the preceding sentence . . ."

The majority of the papers that I have studied have been explained to me by either my students or my friends. I understand better the mathematicians of the last century, especially Poincaré, but I find those of the seventeenth century to be the most clear and basically more modern.'

V. I. Arnol'd

An almost-lost work

Many Greek mathematical works are known to have been lost, recalled only in references by other writers. How many works of European mathematics have also disappeared?

'Desargues' classic work, the one which might have been rightly termed the beginnings of the projective geometry, and which we term PG17, was a treatment of conics having the title *Brouillon projet d'une atteinte aux événemens des recontres d'un cone avec un plan*. Published in 1639, it was a highly original and brilliant work, certainly on a par with any work of his contemporaries. But the quality of a work is not the sole determinant of its acceptance in the macrocosm of learning, and it proved to be, in the words of Boyer, "one of the most unsuccessful great books ever produced" . . . Only around 50 copies were printed and distributed to a selected group. All of these copies disappeared until one was found about 1950 in the Bibliothèque National, Paris, containing notes and errata by Desargues himself. However, fortunately a manuscript copy made by Philippe de La Hire was discovered earlier – in 1845 – so that the later geometers of the 19th century were not entirely deprived of a knowledge of Desargues and his work . . .'

From a Kindergarten Teacher

'A small boy of five came into the kindergarten one morning with radiant face and sparkling eyes, crying out in joyful tones: "I have something for you! It's hard and long and has four edges and two ends!" The precious object was held behind him, while he danced around in fond anticipation of the pleasure he was about to give his teacher, of whom he was very fond. "What can it be?" she answered, entering sympathetically into his pleasure. "Do show it to me." In proud triumph the hand which held the treasure was extended, and in the palm lay a burnt match. And the kindergarten teacher accepted it as a gift of value, for had it not helped to unlock the great world of form and its elements – faces, corners, and edges?'

Cardano wheedles the secret of the cubic out of Tartaglia

'CARDANO: I hold it very dear that you have come now, when his Excellency the Signor Marchese had ridden as far as Vigevano, because we will have the opportunity to talk, and to discuss our affairs together until he returns. Certainly you have, alas, been unkind in not wishing to give me the rule that you discovered, on the case of the thing and the cube equal to a number, even after my greatest entreaties for it.

TARTAGLIA: I tell you, I am not so unforthcoming merely on account of the solution, nor of the things discovered through it, but on account of those things which it is possible to discover through the knowledge of it, for it is a key which opens the way to the ability to investigate boundless other cases. And if it were not that at present I am busy with the translation of Euclid into Italian (and at the moment I have translated as far as his thirteenth book), I would already have found a general rule for many other cases. But as soon as I have completed this work on Euclid that I have already begun, I intend to compose a book on the practice [of arithmetic], and together with it a new algebra, in which I have resolved not only to publish to every man all my discoveries of new cases already mentioned, but many others which I hope to find; and, more, I want to demonstrate the rule that enables one to investigate boundless other cases, which I hope will be a useful and beautiful thing. And this is the reason which makes me refuse them to everyone, because at present I am not working on them (being, as I said, busy with Euclid), and if I teach them to any speculative person (as is your Excellency), he could easily with such clear information find other solutions (it being easy to combine it with the things already discovered), and publish it, as inventor. And to do that would spoil all my plans. Thus this is the principal reason that has made me so unkind to your Excellency, so much more as you are at present having your book printed on a similar subject, and even though you wrote to me that you want to give out these discoveries of mine under my name, acknowledging me as the inventor. Which in effect does not please me on any account, because I want to publish these discoveries of mine in my books, and not in another person's books.

CARDANO: And I also wrote to you that if you did not consent to my publishing them, I would keep them secret.

TARTAGLIA: It is enough that I did not choose to believe that.

CARDANO: I swear to you, by God's holy Gospels, and as a true man of honour, not only never to publish your discoveries, if you teach me them, but I also promise you, and I pledge my faith as a true Christian, to note them down in code, so that after my death no one will be able to understand them. If you want to believe me now, then believe me, if not, leave it be.

TARTAGLIA: If I did not give credit to all your oaths, I would certainly deserve to be judged a faithless man, but since I have decided to ride to Vigevano to call upon his Excellency the Signor Marchese, because it is now three days that I have been here, and I am sorry to have waited for him so long, when I have returned I promise to demonstrate everything to you.

CARDANO: Since you have decided anyway to ride as far as Vigevano after the Signor Marchese, I want to give you a letter to give to his Excellency, so that he should know who you are. But before you go, I want you to show me the rule for these solutions of yours, as you have promised me.

TARTAGLIA: I am satisfied. But I want you to know, that, to enable me to remember the method in any unforeseen circumstance, I have arranged it as a verse in rhyme, because if I had not taken this precaution, I would frequently have forgotten it, and although my telling it in rhyme is not very concise, it has not bothered me, because it is enough that it serves to bring the rule to mind every time that I recite it. And I want to write down this verse for you in my own hand, so that you can be sure that I am giving you the invention accurately and well.'

Cardano (1501–1576) in due course broke his promise to Tartaglia (*c.* 1499–1557) and published the method in his *Ars Magna* (1545).

Hot stuff

Joseph Fourier (1768–1830) was a teacher of mathematics at the prestigious Ecole Polytechnique who accompanied Napoleon on his campaign in Egypt. Fourier is best remembered for his *Analytic Theory of Heat*, once described as 'a great mathematical poem'.

'It seems that [Fourier] from his experience in Egypt, and maybe his work on heat, became convinced that desert heat is the ideal condition for good health. He accordingly clothed himself in many layers of garments and lived in rooms of unbearably high temperature. It has been said that this obsession with heat hastened his death, by heart disease, so that he died, thoroughly cooked, in his sixty-third year.'

Newton = 10 × Dryden

'Leibniz did not begrudge the poet Dryden the sum of £1,000 which he received for his translation of Virgil, but would have liked the astronomer Halley to have received four times as much, and Isaac Newton, ten times as much.'

For Newton's masterpiece, *Principia Mathematica*, which Edmund Halley (1656–1742) was instrumental in bringing to publication.

Dreaming a solution

Leonard Eugene Dickson, number theorist and historian of mathematics, recounted this anecdote of his mother and aunt when they were still children:

'[His] mother and her sister, who, at school, were rivals in geometry, had spent a long and futile evening over a certain problem. During the night, his mother dreamed it and began developing the solution in a loud and clear voice; her sister, hearing that, arose and took notes. On the following morning in class, she happened to have the right solution which Dickson's mother failed to know.'

Ramanujan's flash

'On another occasion, I went to his room to have lunch with him. The First World War had started sometime earlier. I had in my hand a copy of the monthly *Strand Magazine* which at that time used to publish a number of puzzles to be solved by readers. Ramanujan was stirring something in a pan over the fire for our lunch. I was sitting near the table, turning over the pages of the *Magazine*. I got interested in a problem involving a relation between two numbers. I have forgotten the details; but I remember the type of the problem. Two British officers had been billeted in Paris in two different houses in a long street; the door numbers of these houses were related in a special way; the problem was to find out the two numbers. It was not at all difficult. I got the solution in a few minutes by trial and error.

MAHALANOBIS: (In a joking way), Now here is a problem for you.

RAMANUJAN: What problem, tell me. (He went on stirring the pan).

I read out the question from the *Strand Magazine*.

RAMANUJAN: Please take down the solution. (He dictated a continued fraction).

The first term was the solution which I had obtained. Each successive term represented successive solutions for the same type of relation between two numbers, as the number of houses in the street would increase indefinitely. I was amazed.

MAHALANOBIS: Did you get the solution in a flash?

RAMANUJAN: Immediately I heard the problem, it was clear that the solution was obviously a continued fraction; I then thought, "Which continued fraction?" and the answer came to my mind. It was just as simple as this.'

P. C. Mahalanobis

Winston Churchill's vision

'I had a feeling once about Mathematics – that I saw it all. Depth beyond Depth was revealed to me – the Byss and the Abyss. I saw – as one might see the transit of Venus or even the Lord Mayor's Show – a quantity passing through infinity and changing its sign from plus to minus. I saw exactly how it happened and why the tergiversation was inevitable – but it was after dinner and I let it go.'

Poets and mathematicians

'Happy the lot of the pure mathematician. He is judged solely by his peers and the standard is so high that no colleague can ever win a reputation he does not deserve. No cashier writes articles in the *Sunday Times* complaining about the incomprehensibility of Modern Mathematics and comparing it unfavourably with the good old days when mathematicians were content to paper irregularly-shaped rooms or fill bath-tubs with the waste-pipe open. Better still, since engineers and physicists have occasionally been able to put his equations to destructive use, he is given a chair . . .'

W. H. Auden

Auden also wrote: 'When I find myself in the company of scientists,

I feel like a shabby curate who has strayed by mistake into a drawing room full of dukes.'

Turing's test – the imitation game

Computers, mechanical or electronic, can certainly do elementary arithmetic very quickly, and calculate logically – but can they think? Can they, not so much *prove* as *think of*, or *imagine* an original mathematical proposition? Or write a sonnet? This is a portion of Alan Turing's response to such questions.

'I propose to consider the question, "Can machines think?" This should begin with definitions of the meaning of the terms "machine" and "think". The definitions might be framed so as to reflect so far as possible the normal use of the words, but this attitude is dangerous. If the meaning of the words "machine" and "think" are to be found by examining how they are commonly used it is difficult to escape the conclusion that the meaning and the answer to the question, "Can machines think?" is to be sought in a statistical survey such as a Gallup poll. But this is absurd. Instead of attempting such a definition I shall replace the question by another, which is closely related to it and is expressed in relatively unambiguous words.

The new form of the problem can be described in terms of a game which we call the "imitation game". It is played with three people, a man (A), a woman (B), and an interrogator (C) who may be of either sex. The interrogator stays in a room apart from the other two. The object of the game for the interrogator is to determine which of the other two is the man and which is the woman. He knows them by labels X and Y, and at the end of the game he says either "X is A and Y is B" or "X is B and Y is A". The interrogator is allowed to put questions to A and B thus:

C: Will X please tell me the length of his or her hair?

Now suppose X is actually A, then A must answer. It is A's object in the game to try and cause C to make the wrong identification. His answer might therefore be:

"My hair is shingled, and the longest strands are about nine inches long."

In order that tones of voice may not help the interrogator the answers should be written, or better still, typewritten. The ideal arrangement is to have a teleprinter communicating between the two rooms. Alternatively the question and answers can be repeated by an intermediary. The object of the game for the third player (B) is to help the interrogator. The best strategy for her is probably to give truthful answers. She can add such things as "I am the woman, don't listen to him!" to her answers, but it will avail nothing as the man can make similar remarks.

We now ask the question, "What will happen when a machine takes the part of A in this game?" Will the interrogator decide wrongly as often when the game is played like this as he does when the game is played between a man and a woman? These questions replace our original, "Can machines think?"

We may now consider the ground to have been cleared and we are ready to proceed to the debate on our question, "Can machines think?" and the variant of it quoted at the end of the last section. We cannot altogether abandon the original form of the problem, for opinions will differ as to the appropriateness of the substitution and we must at least listen to what has to be said in this connexion.

It will simplify matters for the reader if I explain first my own beliefs in the matter. Consider first the more accurate form of the question. I believe that in about fifty years' time it will be possible to programme computers, with a storage capacity of about 10^9, to make them play the imitation game so well that an average interrogator will not have more than 70 per cent chance of making the right identification after five minutes of questioning. The original question, "Can machines think?" I believe to be too meaningless to deserve discussion. Nevertheless I believe that at the end of the century the use of words and general educated opinion will have altered so much that one will be able to speak of machines thinking without expecting to be contradicted. I believe further that no useful purpose is served by concealing these beliefs. The popular view that scientists proceed inexorably from well-established fact to well-established fact, never being influenced by any improved conjecture, is quite mistaken. Provided it is made clear which are proved facts and which are conjectures, no harm can result. Conjectures are of great importance since they suggest useful lines of research.'

The many proofs of Pythagoras

Elisha Scott Loomis published in 1940 *The Pythagorean Proposition*, a labour of love which contained 370 different proofs, including proofs by Bhaskara, Fibonacci, Leonardo da Vinci, Wallis, Huygens, Leibniz, Saunderson (who was blind yet held Newton's chair at Cambridge, see page 16), De Morgan, and James Garfield, twentieth President of the United States, a blind girl, Miss E. A. Coolidge, and several proofs by teenagers.

This is Leonardo's proof. The thin lines divide the figure into four quadrilaterals, which are congruent to each other, as can be seen by imagining that BA is rotated clockwise about B until it coincides with BX, when quadrilateral BAUY will coincide with quadrilateral BXVC.

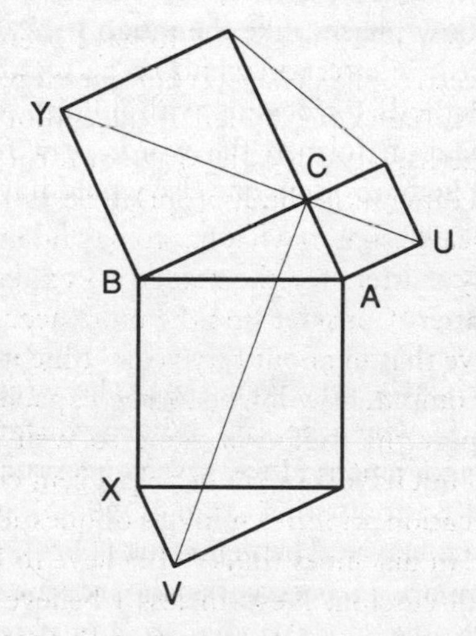

Mathematics and Military Affairs

'As for the knowledge of lines, superficies, and bodies, though it be a science of much certainty and demonstration, it is not much useful for a gentleman, unless it be to understand fortifications, the knowledge whereof is worthy of those who intend the wars; though yet he must remember, that whatsoever art doth in way of defence, art likewise, in way of assailing, can destroy. This study hath cost me much labour, but as yet I could never find how any place could be so fortified, but that there were means in certain opposite lines to prevent or subvert all that could be done in that kind.'

Lord Herbert of Cherbury (1582–1648)

Two Anagramatists

In the days when there were no recognized scientific journals, Galileo and Newton both used anagrams to publish discoveries, and so claim priority, while hiding the actual contents.

Galileo announced his important observation of the phases of Venus in a letter dated 11 December 1611. (Copernicus had predicted that if the planets orbit the sun rather than the earth then Venus should show phases, like the moon.)

Newton wrote in a letter to Leibniz on 24 October 1676, giving details of the methods for dealing with infinite series. The letters of the giant anagram formed the words of a forty-word Latin sentence, which Leibniz or anyone else would have had no chance at all of discovering.

Flaubert teases

'When Flaubert was a very young man, he wrote a letter to his sister, Carolyn, in which he said: "Since you are now studying geometry and trigonometry, I will give you a problem. A ship sails the ocean. It left Boston with a cargo of wool. It grosses 200 tons. It is bound for Le Havre. The mainmast is broken, the cabin boy is on deck, there are 12 passengers aboard, the wind is blowing East-North-East, the clock points to a quarter past three in the afternoon. It is the month of May – How old is the captain?" '

'Seniora Wrangler'

The men in the Cambridge Tripos examinations used to be listed from the Senior Wrangler at the top, downwards. After the first women's colleges, Girton and Newnham, were opened in the 1870s, the women 'wranglers' had their own list on which they were compared to the men, rather than men and women being arranged on the same list.

In 1880 Charlotte Scott was bracketed with the Eighth Wrangler. In 1890, Philippa Fawcett came 'above the Senior Wrangler'. The men didn't accept this, of course, and the title of Senior Wrangler still went to the top man – which prompts the question: how did he feel?

Punch wittily described Philippa Fawcett as 'Seniora Wrangler' on the grounds that she was Seniorer to the Senior Wrangler, and it was her success which inspired Bernard Shaw to make the heroine of his play *Mrs Warren's Profession* a mathematician.

It was more than fifty years later, in 1948, that women were admitted to degrees at Cambridge on equal terms with men.

African river-crossing problems

The earliest appearance of river-crossing puzzles is in the *Propositiones ad Acuendos Juvenes*, or 'Propositions to sharpen the young' traditionally attributed to Alcuin (*c.* 732–804). This is his eighteenth problem.

'A man takes a wolf, a goat and a cabbage across a river. The only boat he could find could take only two of them at a time. But he had been ordered to transfer all of these to the other side in good condition. How could this be done?'

This puzzle, and variants, appear, with the same logical structure, not only in printed sources, but in collections of Gaelic, Danish, Russian, Italian, Romanian and Black American folklore. African versions, however, appear in two distinct logical forms, in the more complicated of which there are extra constraints which make the solution harder.

'The Kpelle problem . . . tells the story of a king who has a caged cheetah that grabs and eats any fowl near it. The king challenges a suitor [*this is another almost universal theme*] for his daughter's hand to transport the cheetah, a fowl and some rice across the river in a boat that holds one person and two of these. But the king points out, the man cannot control them while rowing the boat, and so in addition to the cheetah and the fowl or fowl and rice not being alone together on either shore, they also cannot be together on the boat.'

Query: did these problems arise independently, because they describe situations that are actually realistic in many traditional societies? Or are they all derived from the same source?

The uses of a sextant

Francis Galton (1822–1911), author of *Hereditary Genius*, was a collector of statistics of all kinds, as well as the discoverer of the useful role of fingerprints in criminal investigations, and an enthusiast for eugenics.

On one of his trips into Africa, meeting the Hottentots, he became interested in the reasons why Hottentot women have large buttocks, and determined to measure the vital statistics of one such woman. He feared, however, that such close attentions might lead to misunderstanding of his intentions, so he solved the problem from a distance, literally, by using his surveyor's sextant. He later incorporated his measurements in a scientific paper.

Advice from Paul Halmos

Halmos has mixed feelings about the role of logic in mathematics and the relationship between logic and mathematics.

If you think that your paper is vacuous,
Use the first-order functional calculus.
It then becomes logic,
And, as if by magic,
The obvious is hailed as miraculous.

Mathematics in solitary

Roger Cooper was jailed in Iran in 1985 and spent several years in solitary confinement.

'Between interrogations, always blindfolded and accompanied by slaps and punches when I refused to confess to being a British spy, I tried to find ways of amusing myself without books. I made a backgammon set with dice of bread, and evolved a maths system based on Roman numerals but with an apple pip for zero. Orange pips were units, plum stones were fives, and tens and hundreds were positional. This enabled me to calculate all the prime numbers up to 5,000, which I recorded in dead space where the door opened and could speculate on the anomalies in their occurrence.

It was over 40 years since I had last studied mathematics formally (for the old School Certificate) although, as a journalist writing on economics and a marketing manager in the oil industry, I always had to be reasonably numerate. I began by trying to remember how to solve quadratic equations and prove Pythagoras' theorem, which took days and consumed many sheets of precious paper (purloined from my interrogators or recycled orange wrapping). Before I could reconstruct the classic solution for the square of the hypotenuse proposition I found two solutions of my own, though I am not quite sure they were both valid. Once I got my calculator back I could explore numbers. I found myself fascinated by recurring decimals, especially reciprocals and their multiples. I noticed that if you divide 1 by 7 and 2 by 7 the digits 28571 occur in both, and it soon became clear that the decimal series were really circles of digits totalling one less than the denominator (so 6 digits when 7 is the denominator). Then I discovered that these circles are in fact geometric progressions for which I discovered a formula. There were many other interesting anomalies and curiosities. One party trick I evolved was to be able to convert a decimal series back into a vulgar fraction. For example, if you divide 101 by 103 you get a decimal series (0.9805825 . . .). It goes on for 102 digits before repeating.

The trick is to turn that back into the two numbers. When I at last got some maths books I did find a cumbersome method using quite advanced algebra. I worked out a terribly simple way, using an ordinary pocket calculator.'

All horses are the same colour

Clearly, one horse is the same colour as itself. Assume that the proposition of the title is true for any set of n horses. Then we shall prove that it is true for any set of $n + 1$ horses.

Consider a set of $n + 1$ horses, and take one horse out of the set. This leaves a set of n horses, which by hypothesis are all of the same colour. Now replace the horse removed, and remove a different horse. Once again, we have a set of n horses remaining, all of the same colour. Therefore the set of $n + 1$ horses is the union of two sets of n horses, each of the same colour. Therefore, all $n + 1$ horses in the set are of the same colour. Therefore, by induction, all horses are of the same colour, as was to be proved.

The mathematicians as preacher

Edmund Gunter (1581–1626) invented Gunther's Chain, still used by surveyors, the decimal separator, and the terms *cosine* and *cotangent*, as well as discovering the phenomenon of magnetic variation.

'When he was a Student at Christchurch, it fell to his lott to preach the Passion Sermon, which some old divines that I knew did heare, but 'twas sayd of him then in the University that our Saviour never suffered so much since his Passion as in that sermon, it was such a lamentable one – *Non omnia possumus omnes* [all things are not possible to all men.] The world is much beholding to him for what he hath donne well.'

Advanced maths

'All maths is friteful and means O but if you are a grate brane you hear a tremendous xplosion at about the fifth lesson in trig. This mean that you are through the sound barrier and maths hav become what every keen maths master tell you it can be i.e. a LANGWAGE. in my view this is just another of those whopers which masters tell i mean can you imagine peason and me at brake:

MOLESWORTH: (*taking a hack at the pill*) $x^2 \times y^6 = a$

PEASON: $z^8 - x^3 = b$

MOLESWORTH: (*missing completely*) O, y^{99}!

As you will see it simply will not do i am prepared to believe that a strate line is infinitely protracted go on for ever tho i do not see how even that weed pythagoras can tell. But if maths is a langwage i hav only one comment. It is

$$\sum_{\bullet}^{\infty} \frac{b^x}{x!} = \sum_{\bullet}^{\infty} \frac{e.^{-M} . M^x . e^{tx}}{x!} = e^{-M} \sum_{\bullet}^{\infty} \frac{(Me^t)^x}{x!}$$

I think that setles the mater.'

Giant numbers

The Indians have long had a fascination with large numbers. In the Ramayana, (after 300 BC) the army of Ravana comprised $10^{12} + 10^5 + 36(10^4)$ men. He faced the army of Rama, who commanded,

$10^{10} + 10^{14} + 10^{20} + 10^{24} + 10^{30} + 10^{34} + 10^{40} + 10^{44} + 10^{52} + 10^{57} + 10^{62} + 5.$

For each of these powers of ten, the Indians had a name.

The Indian Jain mathematicians distinguished between enumerable numbers, divided into the lowest, intermediate and highest, the innumerable, divided into nearly innumerable, truly innumerable and innumerably innumerable, and the infinite, classified into nearly infinite, truly infinite and infinitely infinite.

Their conception of the largest enumerable numbers is indicated by this quotation from Anujoga Dwara Sutra, dated about the beginning of the Christian era:

'Consider a trough whose diameter is that of the earth (100,000 yojanna) and whose circumference is 316,227 yojanna. Fill it up with white mustard seeds counting one after the other. Similarly fill up with mustard seeds other troughs of the sizes of the various lands and seas. Still the highest enumerable number has not been attained.'

Harvard in 1802

Traditionalists are fond of recalling a Golden Age when every school pupil could spell, and do simple arithmetic. They should not go *too far back* in search of this Elysium:

The statutes of the University of Prague for the year 1367 required for the M.A. degree, the 'Algorismus', that is, the elements of practical arithmetic. For this, three weeks of lectures were given. The requirements for geometry were also modest. Typically, the M.A. degree demanded only the first five or six books of Euclid, out of thirteen in total.

In nineteenth-century America, standards should have been much higher. Were they? 'In 1802 the standard of admission to Harvard College was raised so as to include a knowledge of arithmetic to the "Rule of Three". A boy could enter the oldest college in America prior to 1803 without a knowledge of a multiplication table.'

In Cambridge, England, in 1829, the standards were even higher. Students read up to the twenty-first proposition of Book 11 and the second proposition of Book 12.

Newton on poets, poets on Newton

A friend once asked Newton, 'Sir Isaac, what is your opinion of poetry?' His answer was: 'I'll tell that of Barrow – he said that poetry was a kind of ingenious nonsense.' Barrow's opinion is consistent with that of the Cambridge mathematician who 'after the perusal of *Paradise Lost* delivered himself of the criticism that the author had proved nothing'.

The poets thought differently of Newton, as did, indeed, everyone. Lagrange summed up a general opinion when he described Newton as 'the greatest genius that ever existed, and the most fortunate, for we cannot find more than once a system of the world to establish'.

James Thomson (1700–1748) was first in the field. Newton died on 20 March 1727, and at the beginning of May 1727 Thomson, who is better known for his *The Seasons*, published 'To the Memory of Sir Isaac Newton':

> O unprofuse magnificence divine!
> O wisdom truly perfect! thus to call
> From a few causes such a scheme of things,
> Effects so various, beautiful, and great,
> An universe complete! And O beloved
> Of Heaven! whose well purged penetrating eye

The mystic veil transpiercing, inly scanned
The rising, moving, wide-established frame.
He, first of men, with awful wing pursued
The comet through the long elliptic curve,
As round innumerous worlds he wound his way,
Till, to the forehead of our evening sky
Returned, the blazing wonder glares anew,
And o'er the trembling nations shakes dismay.
The heavens are all his own, from the wide rule
Of whirling vortices and circling spheres
To their first great simplicity restored.
The schools astonished stood; but found it vain
To combat still with demonstration strong,
And, unawakened, dream beneath the blaze
Of truth. At once their pleasing visions fled,
With the gay shadows of the morning mixed,
When Newton rose, our philosophic sun!

Pope's 'Epitaph Intended for Sir Isaac Newton' is the best-remembered of his lines, and also the subject of parody and respectful imitation. It is sometimes claimed that it was intended to adorn Newton's tomb before the Choir Screen in Westminster Abbey, though the couplet was probably as much satirical as admiring: after all, it was Pope who warned his readers, '. . . attempt not God to span, the proper study of Mankind is Man'. Anyway, it was first published in the *Grub-Street Journal* of 16 July 1730.

Nature and Nature's Laws lay hid in Night:
God said, *Let Newton be!* and all was light.

Pope's lines invite imitation. The great optician Ernst Abbé was honoured with the lines,

Objectives and their laws lay hid in night;
God said, Let Abbé be, and all was light.

The temptation to students of optics is obvious, and fitting, in view of Newton's original researches. In 1867 Helmholtz was praised as the inventor of the ophthalmoscope and founder of optical diagnosis, with these lines:

L'ophthalmologie était dans les ténèbres –
Dieu parla, qui Helmholtz naquit – Et la lumière est faite!

More recently, Sir John Squires capped Pope thus:

It did not last: the Devil howling 'Ho!
Let Einstein be!' restored the *status quo*.

Wordsworth was a student at St John's: his rooms were just to the north of Trinity College, so he could well have seen the Ante-chapel, and Roubiliac's statue, as he describes it in *The Prelude*.

And from my pillow, looking forth by light
Of moon or favouring stars, I could behold
The antechapel where the statue stood
Of Newton with his prism and silent face.
The marble index of a mind for ever
Voyaging through strange seas of Thought, alone.

A canonical position

'Felix Klein compared a high-school pupil with a cannon which is being charged with knowledge for ten years and then fired after which nothing remains.'

John Aubrey on mathematics and small children

'A child of seven years is capable of adding and multiplying and can do as much in this as the profoundest logist. When arithmetic is not this way learnt and made habitual, it makes men almost mad . . . Whereas for addition, or multiplication or division we see little boys run as it were, a gallop, and not make a false step.

Mr Lidcot of Dublin, a learned gentleman, has a daughter of eleven years of age that understands all arithmetic and algebra, trigonometry and the use of globes, and appears at the Royal Society there. They do not find anything extraordinary in her nature to mathematics, but do impute all to her early education . . . Doctor Holder's niece, 6 years old, going on 7, adds, subtracts, multiplies, divides and has some few definitions of geometry. Mr Thomas Ax told me that his father taught him the table of multiplication when he was seven years of age . . .

When the child understands addition, subtraction, multiplication and division (called algorithms) they will quickly learn the rule of three . . . Then let them learn the use of globes . . . Sir Geoffrey Chaucer wrote a treatise for the use of the astrolabe for the use of his son . . . He begins thus:

Little Lovis, my sonne, I perceive well by certain evidences thine abilities to learne sciences, touching numbers and proportions, and also well consider I thy busy prayer in especiall to learne the treatise of the astrolabes. This treatise will I shew thee in wonder light rules and naked words in English, for Latine ne canst thou not yet but small, my little sonne.

It seems by this introduction he was but ten years of age . . .'

Unearthly powers

'Give me a place to stand on and I will move the earth.'

Archimedes

'What Archimedes said of the mechanical powers may be applied to reason and liberty. "Had we," says he, "a place to stand upon, we might raise the world." '

Thomas Paine

The Lanchester model

Frederick Lanchester was a pioneer of automobile design – the "Lanchester" went into production in 1900 – and of the use of aircraft in warfare. In 1916 he published *Aircraft in Warfare, the Dawn of the Fourth Arm*, in which he included an analysis of the fate of opposing forces in battle, according to mathematical measures of their numbers and firepower, concluding that 'the *fighting strength* of a force may be broadly defined as proportional to *the square of its numerical strength multiplied by the fighting value of its individual units*.'

More recently, Martin Braun fitted 'Lanchester's square law' as it is called, to data from the battle of Iwo Jima in World War II. The battle started on 19 February 1945 and lasted a month. The Americans lost nearly 7,000 troops dead, and nearly 20,000 wounded. The Japanese started with approximately 21,000 and fought to the last man, as ordered, losing 20,000 dead. From the detailed data on reinforcement rates, and combat losses, Braun constructed this graph, showing the astonishing fit between the actual American troop strengths during the battle, and the strengths predicted by Lanchester's model:

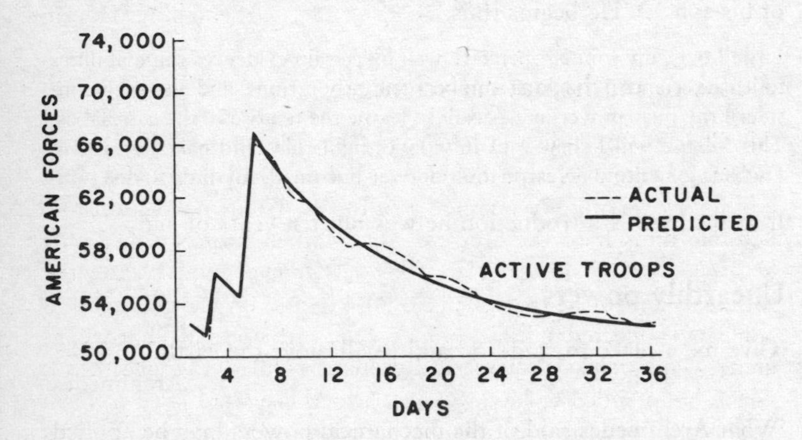

The Immortal Dinner

'On December 28th the immortal dinner came off in my painting-room, with Jerusalem towering up behind us as a background. Wordsworth was in fine cue, and we had a glorious set-to – on Homer, Shakespeare, Milton and Virgil. Lamb got exceedingly merry and exquisitely witty; and his fun in the midst of Wordsworth's solemn intonations of oratory was like the sarcasm and wit of the fool in the intervals of Lear's passion. He made a speech and voted me absent, and made them drink my health. "Now," said Lamb, "you old lake poet, you rascally poet, why do you call Voltaire dull?" We all defended Wordsworth, and affirmed there was a state of mind when Voltaire would be dull. "Well," said Lamb, "here's Voltaire – the Messiah of the French nation, and a very proper one too."

He then, in a strain of humour beyond description, abused me for putting Newton's head into my picture; "a fellow," said he, "who believed nothing unless it was as clear as the three sides of a triangle." And then he and Keats agreed he had destroyed all the poetry of the rainbow by reducing it to the prismatic colours.

It was impossible to resist him, and we all drank "Newton's health, and confusion to mathematics." It was delightful to see the good humour of Wordsworth in giving in to all our frolics without affectation and laughing as heartily as the best of us.'

Benjamin Haydon

A loss to mathematical science

'Sir Charles Cavendish [1591–1654] was the younger Brother to William, Duke of Newcastle. He was a little, weake, crooked man, and nature having not adapted him for the Court nor Campe, he betooke himself to the Study of the Mathematiques, wherin he became a great Master. His father left him a good Estate, the revenue whereof he expended on bookes and on learned men.

He had collected in Italie, France, &c., with no small chardge, as many Manuscript Mathematicall bookes as filled a Hogges-head, which he intended to have printed; which if he had lived to have donne, the growth of Mathematicall Learning had been 30 yeares or more forwarder then 'tis. But he died of the Scurvey, contracted by hard study, about 1652, and left an Attorney of Clifford's Inne, his Executor, who shortly after died, and left his Wife Executrix, who sold this incomparable Collection aforesaid, by weight to the past-board makers for Wast-paper. A good Caution for those that have good MSS. to take care to see them printed in their life-times.'

Blaise Pascal meets Descartes

'In the summer of 1647 Blaise moved to Paris. The fame of his experiments had preceded him and he received many visitors, including Descartes. Descartes had a prejudice against the Pascals [father Etienne and son Blaise] and he refused to recognise the originality of the paper on conic sections which Blaise had written at the age of 16. The story of the visit is told in a letter written by Blaise's sister Jacqueline. Descartes arrived with a retinue of three friends and several small boys. Blaise, who had the support of his friend de Roberval, a difficult and disputatious man, was already an invalid. He was to be the victim of illness for the rest of his life. First Descartes examined the calculating machine and gave his congratulations to the young inventor. "After that, they began to talk about the vacuum. Monsieur Descartes, when he had listened to an account of the experiment, said, in reply to my brother (who asked him what he thought could have got into the tube): 'Why, a subtle form of matter.' My brother answered as best as he could, but M. de Roberval, seeing that my brother fetched his voice with difficulty, began attacking M. Descartes (keeping just within the bounds of politeness), who replied, somewhat sourly, that he did not mind debating with my brother, who was a reasonable man, and would talk with him as long as he liked, but he had nothing to say to M. de Roberval, who, he declared, was influenced by prejudice."'

The lion's claw

'In 1696 John Bernoulli proposed to the mathematical world the famous Problem of Quickest Descent . . . On receiving this challenge Newton promptly solved the problem, the same evening after a busy day's work at the Mint. The receipt of the solution which was sent anonymously to Bernoulli drew forth the reply, "the lion is known by his claw".'

'The problem of quickest descent requires a curve down which a particle can slide most quickly under gravity from a given point A to a given point B.' The point of this problem is that the solution is *not* the straight line joining A to B, because the particle reaches B faster if its path starts by descending steeply, so that it gains speed rapidly at the start of its journey. Newton proved that the quickest path is a cycloid.

An unusual god

'There is a very good saying that if triangles invented a god, they would make him three-sided.'

Montesquieu (1689–1755)

John von Neumann, by his daughter

'He was also profoundly concerned, however, with the nature of his legacy in this world, in two respects. One had to do with the durability of his work, his intellectual contributions; he was surprisingly insecure about that. And, interestingly enough, I don't think he was a very accurate prophet regarding what turns the practical applications of his pioneering work would take . . .

The second focus of John von Neumann's concern about his earthly legacy was, to put it succinctly, me. I was his only offspring and, towards the end of his life, he became acutely conscious that all his eggs were in one basket, genetically speaking . . . So he put tremendous pressure on me to perform up to the peak of my abilities, and made clear his displeasure with the path I appeared to be taking. I had married young, right out of college, and he thought that this was a bad beginning . . . simply because he feared that a woman who married young was very probably reducing her chances of making a significant intellectual or professional contribution . . .

I'm deeply sorry that he never got to know the grandson who has translated a six-year-old's dream of "someday finding a cure for cancer" [von Neumann died of cancer, see page 19] into a career as a molecular biologist doing research on the chemistry of intercellular message transmission . . .

I'm equally sorry that he could never know the granddaughter who, having avoided scientific subjects throughout her university career, realised after graduation that her newly-focused goal of radically improving health care would require medical training . . .

John von Neumann would have felt reassured and gratified, I believe, by the choices his progeny and his progeny's progeny have made, to do what he considered most important, that is, to utilise our intellectual capacities right up to their limits, to try to achieve whatever potential we have.'

Hilbert

'Hilbert in his researches on integral equations, considered infinite sequences that were square summable. Eventually such sequences came to be regarded as "points" in an infinite dimensional space, and an appropriate "geometry" was developed. Although Hilbert did not take this point of view, such spaces and analogous spaces of square integrable functions became known as "Hilbert Spaces". One day Hilbert was attending a mathematical meeting with his colleague Courant. At this meeting it seemed that every other paper referred to this or that Hilbert Space, or this or that property of Hilbert Space. After one of these papers, Hilbert is reported to have turned to Courant and said "Richard, exactly what is a Hilbert Space?"'

Rousseau on geometry

'I never got so far as to understand properly the applications of algebra to geometry. I did not like this method of working without knowing what I was doing; and it appeared to me that solving a geometrical problem by means of equations was like playing a tune by simply turning the handle of a barrel organ.'

'I have said that [deductive Euclidean] geometry is above the capacity of children; but it is our own fault. We do not perceive that their method is not ours; that what is for us the art of reasoning must be for them only the art of seeing. Instead of teaching them our method, we ought to study theirs; for our way of learning geometry is as much an affair of the imagination as of reasoning. When the enunciation is given, we have to *imagine* the proof, that is, we endeavour to find from what proposition already known the other is a consequence, and from all the consequences which may be drawn from such proposition, to choose the precise one which is relevant.

By this method the most exact reasoner, unless he has a gift of invention, will be brought to a standstill. And, what is the result? Instead of teaching us to find the proofs, they dictate them to us; instead of teaching the pupil to reason, the master reasons for him and exercises only his memory.'

Dirac on mathematical beauty

'It is more important to have beauty in one's equations than to have them fit experience . . . It seems that if one is working from the point of view of getting beauty in one's equations, and if one has really a sound insight, one is on a sure line of progress.'

'The foundations of the theory [of General Relativity] are, I believe, stronger than what one could get simply from the support of experimental evidence. The real foundations come from the great beauty of the theory . . . It is the essential beauty of the theory which I feel is the real reason for believing it.'

Mathematical memories

'Any number of instances occur of mathematicians who cannot recall how they were led to some of their results or how they were proved. They have even forgotten their own authorship and have queried, when proposed to them by others, the correctness of results long since proved by themselves. But there are far more glaring instances of forgetfulness. Thus a distinguished professor at Camford University – or was it Oxbridge? – published a paper but was informed afterwards that he had been anticipated both in his result and proof. He found, probably to his great dismay, not only that he had been anticipated, but that he had actually been the referee for the paper when it had been submitted for publication. Mention may be made of a paper I published more than a quarter of a century ago, a copy of which was sent to a well-known European mathematician. About ten years later, he published a practically identical proof. I then sent him another copy of my paper and wrote that he might find it interesting. He replied that it was!'

L. J. Mordell

A strange agreement

'In 1694, Johann Bernoulli and De L'Hôpital had reached an agreement. De L'Hôpital was to pay an annual sum of 300 francs (which was raised some time afterwards). In exchange De L'Hôpital demanded:

1 That Bernoulli was to solve all mathematical problems submitted by De L'Hôpital.
2 That Bernoulli should not disclose his findings to anybody but De L'Hôpital.
3 That Bernoulli should not pass on to Varignon or others copies of his writings done for De L'Hôpital.'

Bernoulli had effectively sold his soul to De L'Hôpital. (Pierre Varignon was another mathematician, now remembered only through Varignon's theorem in mechanics, and Varignon's parallelogram in vector geometry.) The results were unhappy – De L'Hôpital pinched Bernoulli's ideas and passed them off as his own. Only on De L'Hôpital's death ten years later was Bernoulli able to openly accuse him of deception.

Michael – the pocket Calculator

'At 4 ft 2 in Michael Tan is a pocket-sized calculator. While most children of his age are still struggling to learn their times tables. Michael who has just turned ten, is leaving his university lecturers agog with his maths mastery.

At the age of seven years and seven months he became the youngest child in the world to pass A-level mathematics. Just four months later he became the world's youngest undergraduate – studying for a BSc in mathematics at Canterbury University in New Zealand . . .

Four times a week Michael strolls across the lush parkland near his family's modest brick suburban house in Christchurch to attend the university with his mother . . . "You can tell if he has had a good lecture. He'll be saying, 'Oh, Mum, I really enjoyed it,' and skipping all the way home." '

Dr Johnson invents bases

'Sir, allow me to ask you one question. If the church should say to you, "Two and three make ten", what would you do? "Sir," said he, "I should believe it, and I would count like this: one, two, three four, *ten*." I was now fully satisfied.'

James Boswell

Mathematics, art and science

'... let me say something about mathematics or mathematicians in general, for the benefit of the non-mathematicians. Let me interject that I feel truly sorry for them. I think they are missing what I think is the most exciting and rewarding intellectual activity.

Mathematics is often compared to the arts, particularly to music, and it is true that in mathematics as in music talent may flower at an astonishingly early age, though it has to be admitted, earlier in music than in mathematics. In mathematics, aesthetic considerations, beauty, simplicity and elegance are very important as well as truth. If one looks at mathematics as a body of knowledge, I think it definitely can be characterized as a science, but if one looks at the way in which it grows and accumulates, the actual doing of mathematics seems much more to be an art. Mathematics is concerned exclusively with objects and structures which are creations of the mind, although they may be suggested by or patterned on things that are found in the so-called real world. And since it deals only with creations of the mind, it is cumulative in a way that the other sciences are not.

Mathematics does not shed the old substance as new is added, in the way the natural sciences, for instance, will do. The work of Euclid, Apollonius or Archimedes, to mention some Greek mathematicians from antiquity, is as valid today as when it was done more than two millennia ago. But while the content or substance remains, the form in which it is presented is ever changing. What we may refer to as the landscape of mathematics may change profoundly from one generation to another, and even during shorter time spans, fundamental changes may occur.

Atte Selberg

Eureka!

'Archimedes made many and various wonderful discoveries. Of all these the one which I will explain seems to be worked out with infinite skill. Hiero was greatly exalted in the regal power at Syracuse, and after his victories he determined to set up in a certain temple a crown vowed to the immortal gods. He let out the execution as far as the craftsman's wages were concerned, and weighed the gold out to the contractor to an exact amount. At the appointed time the man presented the work finely wrought for the king's acceptance, and appeared to have furnished the weight of the crown to scale.

However, information was laid that gold had been withdrawn, and that the same amount of silver had been added in the making of the crown. Hiero was indignant that he had been made light of, and failing to find a method by which he might detect the theft, asked Archimedes to undertake the investigation. While Archimedes was considering the matter, he happened to go to the baths. When he went down into the bathing pool he observed that the amount of water which flowed outside the pool was equal to the amount of his body that was immersed. Since this fact indicated the method of explaining the case, he did not linger, but moved with delight he leapt out of the pool, and going home naked, cried aloud that he had found exactly what he was seeking. For as he ran he shouted in Greek: eureka, eureka.

Then, following up his discovery, he is said to have taken two masses of the same weight as the crown, one of gold and the other of silver. When he had done this, he filled a large vessel to the brim with water, into which he dropped the mass of silver. The amount of this when let down into the water corresponded to the overflow of water. So he removed the metal and filled in by measure the amount by which the water was diminished, so that it was level with the brim as before. In this way he discovered what weight of silver corresponded to a given measure of water.

After this experiment he then dropped a mass of gold in like manner into the full vessel and removed it. Again he added water by measure, and discovered that there was not so much water; and this corresponded to the lessened quantity [volume] of the same weight of gold compared with the same weight of silver. He then let down the crown itself into the vase after filling the vase with water, and found that more water flowed into the space left by the crown than into the space left by a mass of gold of the same weight. And so from the fact that there was more water in the case of the crown than in the mass of gold, he calculated and detected the mixture of the silver with the gold, and the fraud of the contractor.'

Vitruvius

Poincaré's memory

'Poincaré [1854–1912] could read at incredible speed, and whatever he read became a permanent possession of his mind. In retention and recall he even exceeded the fabulous Euler. He was particularly strong in spatial memory, and could always recall the very page and line of a book where any particular statement had been made. Another of his peculiarities, brought on by his poor eyesight, was his ability to absorb theorems and passages of mathematics chiefly by ear, rather than eye, as most other mathematicians. He developed this ability in school, where, unable to see the blackboard well, he sat back and listened, following and remembering everything without taking any notes.'

The thirteen duels

'The most frequently told story about János Bolyai [1802–1860] concerns the succession of duels he fought with thirteen of his brother officers. As a consequence of some friction, these thirteen officers simultaneously challenged János, who accepted with the proviso that between duels he should be permitted to play a short piece on his violin. The concession granted, he vanquished in turn all thirteen of his opponents. What is seldom told is what happened very shortly after the batch of duels. János was promoted to a captaincy on the condition that he immediately retire with the pension assigned to his new rank. The government felt bound to consult its interests, for it could hardly suffer the possibility of such an event recurring.'

The mysterious source

'The fairest thing we can experience is the mysterious. It is the fundamental emotion which stands at the cradle of true art and science. He who knows it not and can no longer wonder, no longer feel amazement, is as good as dead, a snuffed-out candle.'

Albert Einstein

From googol to googolplex

'Now here is the name of a very large number: "Googol". Most people would say, "A googol is so large that you cannot name it or talk about it; it is so large that it is infinite." Therefore, we shall talk about it, explain exactly what it is, and show that it belongs to the very same family as the number 1.

A googol is this number which one of the children in the kindergarten wrote on the blackboard:

10000000000000000000000000000000000000000000000000
000000000000000000000000000000000000000000000000000.

The definition of a googol is: 1 followed by a hundred zeros. It was decided, after careful mathematical researches in the kindergarten, that the number of raindrops falling on New York in 24 hours, or even in a year or in a century, is much less than a googol. Indeed, the googol is a number just larger than the largest numbers that are used in physics or astronomy.

Words of wisdom are spoken by children at least as often as by scientists. The name "googol" was invented by a child (Dr Kasner's nine-year-old nephew) who was asked to think up a name for a very big number, namely, 1 with a hundred zeros after it. He was very certain that this number was not infinite, and therefore equally certain that it had to have a name. At the same time that he suggested "googol" he gave a name for a still larger number: "Googolplex". A googolplex is much larger than a googol, but is still finite, as the inventor of the name was quick to point out. It was first suggested that a googolplex should be 1, followed by writing zeros until you got tired. This is a description of what would happen if one actually tried to write a googolplex, but different people get tired at different times and it would never do to have Carnera a better mathematician than Dr Einstein, simply because he had more endurance. The googolplex then, is a specific finite number, with so many zeros after the 1 that the number of zeros is a googol. A googolplex is much bigger than a googol, much bigger even than a googol times a goolgol. A googol times a googol would be 1 with 200 zeros, whereas a googolplex is 1 with a googol of zeros. You will get some idea of the size of this very large but finite number from the fact that there would not be enough room to write it, if you went to the farthest star, touring all the nebulae and putting down zeros every inch of the way.'

Dynamic verse

Mathematicians occasionally believe that they can write poetry – Lewis Carroll thought he could – but they usually write only verse, for better or worse, usually the latter. James Clerk Maxwell, the great mathematical physicist, was realistic about his own capacities. He once wrote 'A Problem in Dynamics', which started,

> An inextensible heavy chain
> Lies on a smooth horizontal plain,
> An impulsive force is applied to A,
> Required the initial motion of K.

This, however, is rather serious, and is also the only poem I know which actually requires a diagram, so here instead are two of his lighter inventions.

IN MEMORY OF EDWARD WILSON,

Who repented of what was in his mind to write after section.

Rigid Body (sings).

> Gin a body meet a body
>   Flyin' through the air,
> Gin a body hit a body,
>   Will it fly? and where?
> Ilka impact has its measure,
>   Ne're a ane hae I,
> Yet a' the lads they measure me,
>   Or, at least, they try.
>
> Gin a body meet a body
>   Altogether free,
> How they travel afterwards
>   We do not always see.
> Ilka problem has its method
>   By analytics high;
> For me, I ken na ane o' them,
>   But what the waur am I?

(CATS) CRADLE SONG.

By a Babe in Knots.

Peter the Repeater,
Platted round a platter
Slips of slivered paper,
Basting them with batter.

Flype 'em, slit 'em, twist 'em,
Lop-looped laps of paper;
Setting out the system
By the bones of Neper.

Clear your coil of kinkings
Into perfect plaiting,
Locking loops and linkings
Interpenetrating.

Why should a man benighted,
Beduped, befooled, besotted,
Call knotful knittings plighted,
Not knotty but beknotted?

It's monstrous, horrid, shocking,
Beyond the power of thinking,
Not to know, interlocking
Is no mere form of linking.

But little Jacky Horner
Will teach you what is proper,
So pitch him, in his corner,
Your silver and your copper.

Hero's powerful magic

Hero of Alexandria wrote in the first century AD. In his *Metrica*, only found in 1896, he proved his formula for the area of a triangle in terms of its sides (the area is $\sqrt{(s(s-a)(s-b)(s-c))}$ where s is one half of the perimeter) in a masterpiece of argument which went outside the usual Greek limitation to three dimensions: his area is the square root of a fourth degree product.

He also composed a *Pneumatics* which perfectly illustrates the serious attention he gave to what we might be tempted to dismiss as toys, were we unable to imagine a world in which any kind of technology was scarce, and general principles to explain their operation even rarer. Hero's devices are sufficiently ingenious to impress us even today. Hero's Magic Horse – *An Automaton, the head of which continues attached to the body, after a knife has entered the neck at one side, passed completely through it, and out at the other; the animal will drink immediately after the operation* – is described in its original form in Dunninger's *Complete Encyclopaedia of Magic*, published in 1967.

Our examples, taken from the first English translation, in 1851, of any Greek technological work, show devices 45 and 50 in the *Pneumatics*: a trick of stability which Hero could not possibly have explained scientifically, and his famous Steam Engine, which illustrates Newton's principle that every action has an equal and opposite reaction, as well as being a forerunner of the jet engine.

'45. *A Jet of Steam supporting a Sphere.*

Balls are supported aloft in the following manner. Underneath a cauldron, containing water and closed at the top, a fire is lighted.

From the covering a tube runs upwards, at the extremity of which, and communicating with it, is a hollow hemisphere. If we put a light ball into the hemisphere, it will be found that the steam from the cauldron, rising through the tube, lifts the ball so that it is suspended.

50. *The Steam-Engine.*

Place a cauldron over a fire: a ball shall revolve on a pivot. A fire is lighted under a cauldron, A B, containing water, and covered at the mouth by the lid C D: with this the bent tube E F G communicates, the extremity of the tube being fitted into a hollow ball, H K. Opposite to the extremity G place a pivot, L M, resting on the lid C D; and let the ball contain two bent pipes, communicating with it at the opposite extremities of a diameter, and bent in opposite directions, the bends being at right angles and across the lines F G, L D. As the cauldron gets hot it will be found that the steam, entering the ball through E F G, passes out through the bent tubes towards the lid, and causes the ball to revolve, as in the case of the dancing figures.

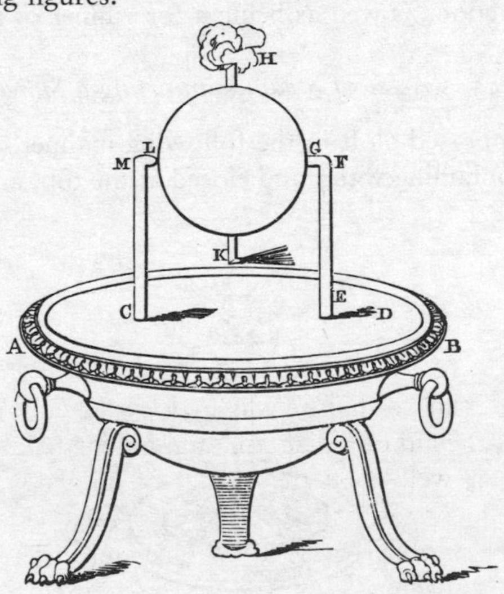

Understanding

John von Neumann had just finished a lecture when a member of the audience came up to him and apologetically explained that he hadn't quite understood von Neumann's concluding argument. Replied von Neumann, 'Young man, in mathematics you don't understand things, you just get used to them.'

Oliver Heaviside, who was a brilliant applied mathematician, made a similar point, in his inimitable way:

'The prevalent idea of mathematical works is that you must understand the reason why first, before you proceed to practise. That is fudge and fiddlesticks. I know mathematical processes that I have used with success for a very long time, of which neither I nor anyone else understands the scholastic logic. I have grown into them, and so understood them that way.'

Brains versus mathematics

'Eddington once introduced some analysis with the words "I regard the introductory part of the theory as the more difficult, because we have to use our brains all the time . . . Afterwards we can use mathematics instead!" '

C. A. Coulson

The death of Archimedes

'But what most of all afflicted Marcellus was the death of Archimedes. For it chanced that he was by himself, working out some problem with the aid of a diagram, and having fixed his thoughts and his eyes as well upon the matter of his study, he was not aware of the incursion of the Romans, or of the capture of the city. Suddenly a soldier came upon him and ordered him to go with him to Marcellus. This Archimedes refused to do until he had worked out his problem and established his demonstration, whereupon the soldier flew into a passion, drew his sword, and dispatched him. However, it is generally agreed that Marcellus was afflicted at his death, and turned away from his slayer as from a polluted person, and sought out the kindred of Archimedes and paid them honour.'

Plutarch

Briggs meets Napier

John Napier, Baron of Marchiston believed that his great gift to mankind was his interpretation of the Apocalypse of St John. We know better. Although now overtaken by electronic calculators, for more than 350 years, logarithms, as it was said, doubled – trebled? – the working life of every astronomer.

'I will acquaint you with a memorable story related to me by Mr John Marr, an excellent mathematician and geometrician whom I conceive you remember, he was a servant to King James, and Charles the First.

At first when Lord Napier, or Lord Marchiston, made publick his Logarithms, Mr Briggs, then reader of the Astronomy lecture at Gresham College in London was so surprized with admiration of them that he could have no quietness in himself, until he had seen that noble person the Lord Marchiston, whose only invention they were. He acquaints John Marr herewith, who goes to Scotland before Mr Briggs, purposely to be there when two so learned persons should meet. Mr Briggs appoints a certain day when to meet in Edinburgh, but failing thereof, the Lord Napier was doubtful he would not come. It happened one day as John Marr and Lord Napier were speaking of Mr Briggs "Ah John", saith Marchiston, "Mr Briggs will not come." At the very instant one knocks at the gate. John Marr hastened down, and it proved Mr Briggs, to his great contentment. He brings Mr Briggs to my Lord's chamber, where almost one quarter of an hour was spent each beholding the other with admiration, before one spoke: at last Mr Briggs began: "My Lord, I have undertaken this long journey purposely to see your person and to know by what engine of wit or ingenuity, you first came to think of this most excellent help unto astronomy, viz. the Logarithms? But my Lord, being by you found out, I wonder nobody else ever found it before, when now, being known, it appears so Easy." He was nobly entertained by Lord Napier, and every summer after this, during Lord Napier's being alive, this venerable man, Mr Briggs, went to Scotland to visit him.'

The End of Mathematics is Nigh

Lagrange (1736–1813) managed to persuade himself that mathematics as a whole was near exhaustion. Charles Babbage asserted in 1813 that, 'The golden age of mathematical literature is undoubtedly past.'

Not so: in 1988, Keith Devlin published the appropriately titled *Mathematics: the New Golden Age*, which discusses, among other topics, new methods of factoring prime numbers and their relation to secret codes, the class number problem, chaos, the classification of the simple groups, the solution of Hilbert's Tenth Problem and the solution by computer of the four-colour conjecture, the efficiency of algorithms and some knotty problems about knots. Only two of these topics were even hinted at in the mathematics of the early nineteenth century. Computers, however, in which Babbage was such a pioneer, are intimately linked to at least four of these themes.

'Let no one ignorant of geometry enter here'

The motto written over Plato's Academy, described by the historian of mathematics, David Eugene Smith, as the world's first recorded college entrance requirement.

Paul Erdös makes an offer

'For Paul Erdös: Among his several greatnesses are an ability to ask the right question and to ask it of the right person.'

Richard Guy

Paul Erdös has spent his entire professional life flitting from one mathematics department to another, meeting other mathematicians, often collaborating with them, and solving problems. He has a charming custom of offering other mathematicians sums of money, from a few dollars to a $1,000 or more, for the solutions to particular problems.

'I also want to mention a very nice problem of Tarski which should be settled: squaring the circle. Can a square and a circle of the same area be decomposed into a finite number of congruent parts? This is a very beautiful problem, and rather well known. If it were my problem I would offer $1,000 for it – a very very nice question, possibly very difficult. Really one has no obvious method of attack. In higher dimensions this is no longer true. As everybody knows, this is the famous Banach–Tarski paradox, the basic idea of which really goes back to Hausdorff in his 1914 book.'

Hardy on proof, and Ramanujan

Ramanujan had an incomparable intuition for correct results, and an amazing capacity to draw conclusions from actual data; and yet his inductions were not always correct, and, as Littlewood remarked, it is unlikely that Ramanujan ever really understood the modern concept of mathematical proof. Hence Hardy's comments:

'Here I am obliged to interpolate some remarks on a very difficult subject: *proof* and its importance in mathematics. All physicists, and a good many quite respectable mathematicians, are contemptuous about proof. I have heard Professor Eddington, for example, maintain that proof, as pure mathematicians understand it, is really quite uninteresting and unimportant, and that no one who is really certain that he has found something good should waste his time looking for a proof.

Ramanujan found the form of the Prime Number Theorem for himself. This was a considerable achievement; for the men who had found the form of the theorem before him, like Legendre, Gauss, and Dirichlet, had all been very great mathematicians; and Ramanujan found other formulae which lie still further below the surface . . .

The fact remains that hardly any of Ramanujan's work in this field had any permanent value. The analytic theory of numbers is one of those exceptional branches of mathematics in which proof really is everything and nothing short of absolute rigour counts. The achievement of the mathematicians who found the Prime Number Theorem was quite a small thing compared with that of those who found the proof. It is not merely that in this theory . . . you can never be sure of the facts without the proof, though this is important enough. The whole history of the Prime Number Theorem, and the other big theorems of the subject, shows that you cannot reach any real understanding of the structure and meaning of the theory, or have any sound instincts to guide you in further research, until you have mastered the proofs. It is comparatively easy to make clever guesses; indeed there are theorems, like "Goldbach's Theorem", which have never been proved and which any fool could have guessed.

The theory of primes depends upon the properties of Riemann's function $\zeta(s)$ considered as an analytic function of the complex variable s, and in particular on the distribution of its zeros; and Ramanujan knew nothing at all about the theory of analytic functions.'

A complex question

'There were innumerable stories about me as an examiner, most of them pure invention. The most famous one happens to be true and it is perhaps worth recording. The candidate shall remain anonymous. He was not terribly good – in mathematics at least. After he had failed to answer a couple of questions, I asked him a really simple one, which was to describe the behavior of the function $1/z$ in the complex plane. "The function is analytic, sir, in the whole plane except at $z = 0$, where it has a singularity," he answered and it was perfectly correct. "What is the singularity called?" I continued. The student stopped in his tracks. "Look at me," I said. "What am I?" His face lit up. "A simple Pole, sir," which is in fact the correct answer.'

Mark Kac

Sign of the Times

Under the heading 'Unsung gay hero of a secret war is honoured with road sign', Martin Whitfield reported recently:

'The great contribution to twentieth-century science of Alan Turing, mathematician, wartime code-breaker and designer of the first electronic computer, is about to achieve belated – if limited – public recognition.

Nearly 40 years after his death, Manchester is to name a section of its new ring road after the homosexual scientist.

If it seems a poor honour for one of the great minds of the age, at least it is something. There were no official awards for Dr Turing, who led the team that read the German Enigma codes in the Second World War. He committed suicide in 1954 after being prosecuted for indecency when aged 41 and working at Manchester University.

After the war, Dr Turing again became involved with the security services but his work at the GCHQ telecommunications centre came to an end when he became a victim of the anti-gay atmosphere of the 1950s. Despite his war record, he was not permitted to visit the United States after his homosexuality was revealed.

In 1938 he was recruited to the cypher school at Bletchley Park, Buckinghamshire, called "Britain's secret weapon" by Churchill. The breaking of the Enigma cyphers there enabled the Allies to anticipate many German military and naval operations, including U-boat attacks.

From 1948, he worked on the Manchester Automatic Digital Machine, the largest computer in the world at that time.

According to friends he was open about his homosexuality, although not ostentatious. He was forced to receive hormone treatment as a condition of being given probation on pleading guilty to charges of gross indecency. The drugs, designed to dampen his libido, made him grow breasts. He committed suicide by eating an apple dipped in cyanide.

Dr Turing was the hero of *Breaking the Code*, a west-end play by Hugh Whitemore in 1986. Graham Stringer, leader of Manchester City Council, said: "Alan Turing has never received the recognition to which he is entitled. We now have a chance to put that right." '

Bertrand Russell dreams his fate

'I can remember Bertrand Russell telling me of a horrible dream. He was in the top floor of the University Library, about AD 2100. A library assistant was going round the shelves carrying an enormous bucket, taking down book after book, glancing at them, restoring them to the shelves or dumping them into the bucket. At last he came to three large volumes which Russell could recognize as the last surviving copy of *Principia mathematica*. He took down one of the volumes, turned over a few pages, seemed puzzled for a moment by the curious symbolism, closed the volume, balanced it in his hand and hesitated . . .'

G. H. Hardy

A surfeit of eggs

'My first visit to Paris was made in company with my friend John Herschel. On reaching Abbeville, we wanted breakfast, and I undertook to order it. Each of us usually required a couple of eggs. I preferred having mine moderately boiled, but my friend required his to be boiled quite hard. Having explained this matter to the waiter, I concluded by instructing him that each of us required two eggs thus cooked, concluding my order with the words, "pour chacun deux".

The garçon ran along the passage half way towards the kitchen, and then called out in his loudest tone –

"Il faut faire bouillir cinquante-deux oeufs pour Messieurs les Anglais." I burst into such a fit of uncontrollable laughter at this absurd misunderstanding of *chacun deux*, for *cinquante-deux*, that it was some time before I could explain it to Herschel, and but for his running into the kitchen to countermand it, the half hundred of eggs would have assuredly been simmering over the fire.

A few days after our arrival in Paris, we dined with Laplace, where we met a large party, most of whom were members of the Institut. The story had already arrived at Paris, having rapidly passed through several editions.

To my great amusement, one of the party told the company that, a few days before, two young Englishmen being at Abbeville, had ordered fifty-two eggs to be boiled for their breakfast, and that they ate up every one of them, as well as a large pie which was put before them.

My next neighbour at dinner asked me if I thought it probable. I replied, that there was no absurdity a young Englishman would not occasionally commit.'

Charles Babbage

Boarding House Geometry

Definitions and Axioms

All boarding houses are the same boarding-house.

Boarders in the same boarding-house and on the same flat are equal to one another.

A single room is that which has no parts and no magnitude.

The landlady of a boarding-house is a parallelogram – that is an oblong angular figure, which cannot be described, but which is equal to anything.

A wrangle is the disinclination for each other of two boarders that meet together but are not in the same line.

All the other rooms being taken, a single room is said to be a double room.

Postulates and Propositions

A pie may be produced any number of times.

The landlady can be reduced to her lowest terms by a series of propositions.

A bee-line may be made from any boarding house to any other boarding house.

The clothes of a boarding-house bed, though produced ever so far both ways, will not meet.

Any two meals at a boarding-house are together less than two square meals.

If from the opposite ends of a boarding-house a line be drawn passing through all rooms in turn, then the stove-pipe which warms the boarders will lie within that line.

On the same bill and on the same side of it there should not be two charges for the same thing.

If there be two boarders on the same flat, and the amount of side of the one be equal to the amount of side of the other, each to each, and the wrangle between one boarder and the landlady be equal to the wrangle between the landlady and the other, then shall the weekly bills of the two boarders be equal also, each to each.

For if not, let one bill be the greater.

Then the other bill is less than it might have been – which is absurd.

Stephen Leacock

A man of figures

'There was Major Hunter, a haunted little man of figures, a little man who, being a dependable unit, considered all other men either as dependable units or as unfit to live. Major Hunter was an engineer, and except in case of war no one would have thought of giving him command of men. For Major Hunter set his men in rows like figures and he added and subtracted and multiplied them. He was an arithmetician rather than a mathematician. None of the humour, the music, or the mysticism of higher mathematics ever entered his head. Men might vary in height or weight or colour, just as 6 is different from 8, but there was little other difference. He had been married several times and he did not know why his wives became very nervous before they left him.'

John Steinbeck

No flies on von Neumann

The following puzzle was put to von Neumann (1903–1957), who had a reputation for calculating fabulously quickly: two cyclists are cycling towards each other, starting twenty miles apart, and cycling at a steady 10 m.p.h. A fly meanwhile starts from the nose of one cyclist, and flies at a steady 15 m.p.h. to the nose of the other cyclist, and then back to the first, and so on, until the cyclists meet. How far does the fly travel? Von Neumann gave the correct solution instantly, whereupon the poser said, in a disappointed tone of voice, 'Oh, you know the trick!' Replied von Neumann, 'What trick? I just added up the infinite series!'

Hard-ly obvious

G. H. Hardy was giving a lecture, and reached a point at which he announced that 'it is now obvious that . . .' Here he paused, and then excused himself and left the room. Twenty minutes later he returned, and repeated triumphantly, 'it *is* now obvious that . . .'

This anecdote has been told of several mathematicians – so there must be some deeper truth in it!

Scaling the peaks

Mathematicians are like mountaineers, they attempt peaks that they judge to be within their reach, just. Tackle problems that are too easy, and you end up dissatisfied with yourself, and, maybe, a damaged career. Tackle problems that are too hard, and potentially fruitful years of research can turn into a desert.

Andrew Wiles (page 203) showed a fine and delicate judgement of his own abilities, when he decided to spend several – seven as it turned out – of the best years of his mathematical life, attempting to crack Fermat's Last Theorem. He succeeded, just.

Excelsior!

Peter Popponesset was as brilliant as could be,
He wouldn't work on any petty problems, no, not he!
His prof suggested topics; Peter said they were too soft,
For he would hold the flag of great significance aloft.
He scorned Riemann's conjecture – he claimed it was passé:
He wouldn't waste his time on classic stumbling blocks today.
He's watching and he's waiting for a challenge to arise
Whose answer will provide sufficient glory in his eyes.
His classmates were pedestrian; they got their Ph.D.'s;
They solved some open problems, and promotions came with ease.
They've even won some prizes given by the A.M.S.,
But Peter sneers at all of them and mocks at their 'success'.
He knows that he'll laugh last, not now, but in the end,
Because he's still expecting inspiration to descend.

Ralph P. Boas

Notes

The mathematician, the physicist and the engineer, page 1

Based on N. J. Rose (ed.), *Mathematical Maxims and Minims*, Rome Press, Raleigh, North Carolina, 1988, p. 128.

Freeman J. Dyson, 'Innovation in Physics', in *Scientific American*, September 1958, pp. 74–82.

On the shores of the unknown, page 1

Aristippus: see frontispiece to J. F. Scott, *A History of Mathematics*, Taylor & Francis, 1960.

Jules Verne, *From the Earth to the Moon*. Note that, according to Jules Verne, the French named as the Ass's Bridge the theorem of Pythagoras, which is much further advanced in Euclid's Elements than the theorem known in England as the Ass's Bridge. See p. 196.

Eddington, quoted in Morris Kline, *Mathematics and the Search for Knowledge*, OUP, 1985, p. 217.

A tricky choice, page 2

Daniel George, *An Eclectic ABC*, Barrie and Rockliff, 1964, pp. 126–7.

Hilbert's memory, page 3

H. Eves, *Mathematical Circles Adieu*, Prindle, Weber and Schmidt, 1977, p. 53.

Conway's Last Theorem, page 3

J. H. Conway, *On Numbers and Games*, Academic Press, 1976, p. 224.

Joe Miller's Jest, page 3

Joe Miller's Jests: or, the Wit's Vade-Mecum, London, 1739, no. 234.

What is mathematics made of?, page 4

G. H. Hardy, *A Mathematician's Apology*, CUP, 1969, p. 84.

H. Aarsleff, *From Locke to Saussure*, Athlone Press, 1982, p. 270.

M. Kline, *Mathematics and the Search for Knowledge*, OUP, 1985, p. 216.

Why ten?, page 4

Aristotle, *Problemata*, Book XV, sec. 3, in *The Works of Aristotle*, trans. W. D. Ross, vol. VII, Clarendon Press, Oxford, 1927.

The largest known prime number, page 5

The Guinness Book of Records, 1994, p. 77.

News Release from California State University, 14 November 1978.

Chrystal clear, page 6

H. Eves, *Mathematical Circles Squared*, Prindle, Weber and Schmidt, 1972, p. 94.

How to be a Good Lecturer, page 7

Jonathan Partington, 'How to be a Good Lecturer', in *Eureka*, 53, February 1994.

Louis Pósa, by Paul Erdös, page 8

Ross Honsberger, *Mathematical Gems*, Mathematical Association of America, 1973, p. 10.

The flies on Fliess, page 9

M. Gardner, *Mathematical Carnival*, Penguin, 1975, pp. 150–51, 152–3, 155. Fliess, being no mathematician, did not realize that even if only positive multiples are allowed, then x and y can still express every integer greater than $xy - x - y$, a result due to Sylvester.

What is a proof?, page 10

Richard Hamming, *Coding and Information Theory*, Prentice-Hall, 1980, p. 155.

From Cicero to typewriters, page 10

Cicero, *De Senectute, De Amicitia, De Divinatione*, trans. W. A. Falconer, Loeb Classical Library, William Heinemann, 1930, Book 1, sec. 23, pp. 249–51.

Bertrand Russell, *Nightmares of Eminent Persons*, quoted in J. Bibby (ed.), *Quotes, Damned Quotes and . . .*, Demast Books, Halifax UK, 1983, p. 38.

E. Kasner and J. Newman, *Mathematics and the Imagination*, Bell, 1970, p. 223.

C. Fadiman (ed.), *Fantasia Mathematica*, Simon and Schuster, 1958, p. 282.

Barrow's humour, page 13

H. Eves, *An Introduction to the History of Mathematics*, Rinehart, New York, 1956, p. 329. Also, entry, 'Barrow', in Clifton Fadiman (ed.), *The Faber Book of Anecdotes*, Faber and Faber, 1985.

Sylvester and *The Laws of Verse*, page 14

Alexander Macfarlane, *Lectures on Ten British Mathematicians of the Nineteenth Century*, Chapman & Hall, 1916, pp. 117–18.

The peripatetic Paul Erdös, page 14

A. Baker, preface to A. Baker, B. Bollobas, A. Hajnal (eds.), *A Tribute to Paul Erdös*, Cambridge University Press, 1990.

John Tierney, 'Paul Erdös is in town. His brain is open', *Science*, 1984, pp. 40–47.

Poincaré's sudden illumination, page 16

H. Poincaré, *Science and Method*, Nelson, 1914, quoted in M. L. Cartwright, *The Mathematical Mind*, OUP, 1955, p. 19.

Saunderson, a blind mathematician, page 16

D. Diderot, *Letter on the Blind*, quoted in M. J. Morgan, *Molyneux's Question*, CUP, 1977, pp. 42–9.

De Moivre approaches the limit, page 18

Clifton Fadiman (ed.), *The Faber Book of Anecdotes*, Faber and Faber, 1985.

Mathematicians in the sky, page 19

Adrian Room, *Dictionary of Astronomical Names*, Routledge, 1988.

The death of John von Neumann, page 19

S. J. Heims, *John von Neumann and Norbert Wiener: From Mathematics to the Technologies of Life and Death*, MIT Press, 1980, pp. 369–71.

Einstein's regrets, page 21

Quoted by S. J. Heims, *John von Neumann and Norbert Wiener: From Mathematics to the Technologies of Life and Death*, MIT Press, 1980, p. 116.

A free slice of pi, page 21

P. Beckmann, *A History of* π, The Golem Press, Boulder, Colorado, 1977, pp. 174, 177.

Pure mathematics, page 22

The Argus, 28 August 1974, quoted in *Mathematical Digest*, 17, October 1974, p. 13.

The Professor, page 22

G. Polya, *How to Solve It*, 2nd edn, Doubleday, New York, 1957, pp. 208–9.

Reason versus Imagination, page 23

Coleridge quote from *Mathematical Digest*, ed. John Webb, 38, January 1980, epigraph.

Martin Gardner's Sixth Book of Mathematical Games from Scientific American, W. H. Freeman, n.d., p. 195.

Gulliver in Lagado, page 23

Jonathan Swift, *Gulliver's Travels and Other Writings*, ed. M. K. Starkman, Bantam Books, New York, 1962, pp. 161, 184.

A notable failure, page 25

From David Singmaster.

God, page 25

The claim is not found in Plato's works, but Plutarch discusses it as Plato's opinion in his *Convivialum disputationem*, 8, 2.

Brigg's epitaph is given in O. L. Dick (ed.), *Aubrey's Brief Lives*, Secker and Warburg, 1949, p. 38.

An Arabian first, page 25

T. H. O'Beirne, *Puzzles and Paradoxes*, OUP, 1965, p. 199 (for text).

O. Ore, *Number Theory and its History*, McGraw-Hill, 1948, p. 140 (for equations).

The Importance of Form, page 26

Anthony Storr, *The Dynamics of Creation*, Penguin, 1976, p. 237, quoting H. G. Gough, 'Identifying the Creative Man', in *Journal of Value Engineering*, vol. 2, 4, 1964, pp. 5–12.

J. E. Littlewood, *A Mathematician's Miscellany*, Methuen, 1963, p. 1.

Maxwell's analogies, page 27

L. C. Woods, 'Maxwell's Models', in *Bulletin of the Institute of Mathematics and its Applications*, vol. 16, 1980, pp. 11–15.

A Valentine, page 27

Eureka, 20, October 1957, p. 18.

A moral tale, page 28

Traditional, or rapidly becoming so.

The landscape of mathematics, page 28

Arthur Cayley, 'Presidential address to the British Association, September 1883', in *The Collected Mathematical Papers of Arthur Cayley*, CUP, 1896, vol. XI, p. 449.

Oliver Wendell Holmes on mathematics, page 28

Oliver Wendell Holmes, *The Autocrat of the Breakfast Table*, sec. 1, E. A. Lawson, New York, n.d. Previously published in *The Atlantic Monthly*, Nov. 1857–Oct. 1858.

The roots of music, page 29

Norman Lebrecht, 'Nobody plays the maestro's way', in *The Sunday Times*, 12 June 1988.

A source of disappointment, page 29

In *Eureka*, 48, March 1988.

Moser and Mordell, page 29

L. J. Mordell, 'Reminiscences of an octogenarian mathematician', in *American Mathematical Monthly*, vol. 78, November 1971.

Julia Robinson and Hilbert's Tenth Problem, page 30

D. J. Albers, G. L. Alexanderson and C. Reid (eds.), *More Mathematical People*, Harcourt, Brace, Jovanovich, 1990, pp. 266–7.

From numbers to God, page 31

Quoted in F. N. David, *Games, Gods and Gambling*, Griffin, 1969, p. 103.

Ramanujan's – what's the word?, page 31

S. R. Ranganathan (ed.), *Ramanujan, the Man and the Mathematician*, Asia Publishing House, 1967, p. 87, recording the reminiscences of T. K. Rajagopalan, and p. 88, recording the reminiscences of R. Srinivasan.

God's Marvelous Mathematical Book, page 32

Quoted in Alexander Soifer, 'From problems of mathematical Olympiads to open problems of mathematics', in *Mathematics Competitions*, vol. 4, 1, 1991, p. 63.

The most prolific mathematician of all time, page 33

C. Truesdell, 'Leonard Euler, Supreme Geometer (1707–1783)', in H. E. Pagliaro (ed.), *Irrationalism in the Eighteenth Century*, Case Western Reserve University Press, 1972, p. 53.

Modern Mathematics for T. C. Mits, The Celebrated Man in the Street, page 34

Lilian R. Lieber, *Modern Mathematics for T. C. Mits, The Celebrated Man in the Street*, George Allen & Unwin, 1946, pp. 77–80.

Dirac on pretty mathematics, page 36

P. A. M. Dirac, 'Pretty Mathematics', in *International Journal of Theoretical Physics*, vol. 21, 8/9, 1982.

A Sampler of Statistics, page 37

John Bibby (ed.), *Quotes, damned quotes, and . . .*, 2nd edn, John Bibby (Books), Edinburgh, 1986.

Counting-out rhymes, page 38

Iona and Peter Opie, *Children's Games in Street and Playground*, OUP, 1984, pp. 36, 47.

J. H. Webb, *Mathematics on Vacation*, University of Capetown, 1987, p. 25.

Melvyn Bragg, *A Place in England*, Secker and Warburg, 1970.

The limits of imagination, page 40

Joseph Addison, 'On the pleasures of the imagination', in the *Spectator*, 420, 2 July 1712.

The pattern of mathematics, page 40

'Matters Mathematical', in *Mathematical Digest*, 57, October 1984, p. 28.

Asymptotes, page 42

F. Le Lionnais, 'Beauty in Mathematics', in *Great Currents of Mathematical Thought*, vol. 2.

Chilling mathematics, page 42

Lancelot Hogben, *Mathematics for the Million*, quoted in Huntley, *The Divine Proportion*, Dover, p. 74.

Saunders MacLane, *Mathematics, Form and Function*, Springer, 1986, p. 456.

Serge Lang, *The Beauty of Mathematics*, Springer, 1985.

Quoted H. R. Pagels, *Perfect Symmetry: the Search for the Beginning of Time*, Penguin, 1992.

B. Russell, 'The Study of Mathematics', in *Philosophical Essays*, quoted R. E. Moritz, *On Mathematics*, Dover, 1942, p. 182.

Why mathematicians exist, page 43

Prof. E. W. Hobson, addressing the British Association for the Advancement of Science, 1910, Section A, Mathematical and Physical Sciences.

A comedy of errors, page 43

G. A. Miller, 'Errors in the Literature of Groups of Finite Order', in *American Mathematical Monthly*, vol. 20, 1913, pp. 14–20.

Gauss's precocity, page 44

H. Eves, *Introduction to the History of Mathematics*, Holt, Rinehart and Winston, 1976, pp. 370–71.

Professional disagreements, page 44

C. Boyer, *Scripta Mathematica*, XVII, 3–4, p. 217.

R. Taton, *Reason and Chance in Scientific Discovery*, Hutchinson, 1957, pp. 30–31.

The Mathematical theory of big game hunting, page 45

H. Eves, *Mathematical Circles Revisited*, P.W.S.-Kent Publishing Co., 1972, pp. 77-8.

H. Petard, 'A contribution to the mathematical theory of big game hunting', in *American Mathematical Monthly*, vol. 45, 446, 1938.

John Barrington, '15 new ways to catch a lion', in I. Stewart and J. Jaworski (eds.), *Seven Years of Manifold 1968–1980*, Shiva Publishing, 1981, pp. 36–9.

Mozart's Musical Game, page 47

M. Eigen and R. Winkler, *Laws of the Game*, Penguin, 1981, pp. 325–7.

Nursery Verses, page 48

Iona and Peter Opie (eds.), *The Puffin Book of Nursery Rhymes*, Penguin, 1966.

Iona and Peter Opie (eds.), *The Oxford Nursery Rhyme Book*, OUP, 1979.

Grace Hopper, page 49

'Grace M. Hopper, 1906–1992', in *Mathematical Digest*, 88, July 1992, p. 17.

The smallest uninteresting number, page 50

Simon Hoggart, the *Guardian*, 25 May 1994.

H. Steinhaus, *Mathematical Snapshots*, OUP, New York, 1969, p. 257.

Freeman Dyson versus Bourbaki, page 50

Freeman H. Dyson, 'Unfashionable Pursuits', in *The Mathematical Intelligencer*, vol. 5, 3, 1983. See p. 206 for Bourbaki.

From amoeba to mathematician, page 50

Lytton Strachey, 'Hume', in *Portraits in Miniature and Other Essays*, Chatto and Windus, 1931.

Babbage corrects Tennyson, page 51

Quoted in Clifton Fadiman, *Fantasia Mathematica*, Simon and Schuster, New York, 1958, p. 293.

Sylvester and Huxley, and structure, page 51

J. Sylvester speaking at the 1869 meeting of The British Association for the Advancement of Science, quoted by P. C. Ritterbush, 'Organic Form: aesthetics and objectivity in the study of form in the life sciences', in *Organic Form, the Life of an Idea*, ed. G. S. Rousseau, Routledge and Kegan Paul, 1972, p. 46.

An IQ test, page 52

R. Falk, 'Magic possibilities of the weighted average', in *Mathematics Magazine*, vol. 53, 2, March 1980.

A popular innovation, page 53

Note 265, *The Mathematical Gazette*, vol. 12, 173, December 1924, p. 232.

Hadamard's failures, page 53

J. Hadamard, *The Psychology of Invention in the Mathematical Field*, Dover, 1954, pp. 51–3.

Omnipresence, page 55

A. Rodin, *Omnibus*, BBC 1, 1986.

Descartes' confidence, page 55

Descartes, *Discourse on Method*, quoted from E. Anscombe and P. T. Geach (eds.), *Descartes: Philosophical Writings*, Open University Press, 1954, pp. 21–2.

The grammar of art, page 56

G. Apollinaire, *Soirées de Paris*, 1912, quoted in 'Malevich's Suprematism', in *The Burlington Magazine*, vol. 118, August 1976, p. 579.

Mathematical medicine, page 56

F. Bowman, *The Mathematical Gazette*, vol. 19, 19, 1935, p. 273.

Leonardo on Mathematics, page 56

The first two quotes are from J. P. Richter (trans.), *The Notebooks of Leonardo da Vinci*, 1970, quoted in H. Kearney, *Origins of the Scientific Revolution*, Longman, 1964, pp. 115–16.

G. A. Tokaty, *A History and Philosophy of Fluidmechanics*, Foulis, Henley-on-Thames, 1971, p. 33.

A. Blunt, *Artistic Theory in Italy*, Clarendon Press, Oxford, 1963, p. 36*n*.

Sonya Kovalevskaya (1850–1891), page 57

Anna Carlotta Leffler, 'Memoir', in Sonia Kovalevsky, *Biography and Autobiography*, Walter Scott, 1895, pp. 34–5, 163–4.

The craft of geometry, page 58

Quoted in F. N. David, *Games, Gods and Gambling*, Griffin, 1969, p. 97.

Nicole Lepaut and the return of Halley's comet, page 59

Carl Sagan and Ann Druyan, *Comet*, Michael Joseph, 1985, pp. 72–5.

The dangers of pattern spotting, page 61

J. E. Littlewood, 'Review of Ramanujan's Collected Papers', in *A Mathematician's Miscellany*, Methuen, 1963, pp. 87–8.

Y. Mikami, *The Development of Mathematics in China and Japan*, Chelsea Publ. Co., New York, 1961, pp. 166, 240.

Vision, page 61

William Blake, 'Auguries of Innocence'.

Alonzo Church, page 62

Gian-Carlo Rota, 'Fine Hall in its golden era', in *A Century of Mathematics in America*, American Mathematical Society, 1989, pp. 224–5.

Mondrian versus the computer, page 63

J. Cohen, *Homo Psychologicus*, George Allen and Unwin, 1970, p. 72. The original experiment was reported in *Cybernetic Serendipity*, Institute of Contemporary Arts, London, 1968.

The conviction of induction, page 64

MacMahon is quoted by G. Polya, *Induction and Analogy in Mathematics*, Princeton University Press, 1954, p. 96.

Richard Guy, 'The Strong Law of Small Numbers', in *American Mathematical Monthly*, vol. 98, 8, 1988, pp. 697–712 and 'The Second Strong Law of Small Numbers', in *Mathematics Magazine*, vol. 63, 1, February 1990, pp. 3–20.

Dr Johnson's advice, page 65

James Boswell, on Dr Samuel Johnson, quoted in *The Mathematical Gazette*, vol. 14, 195, July 1928, p. 156.

Diderot at Court, page 65

Entry, 'Diderot' in Clifton Fadiman (ed.), *The Faber Book of Anecdotes*, Faber and Faber, 1985.

The wrong brother, page 66

E. T. Bell, quoted in Clifton Fadiman (ed.), *The Faber Book of Anecdotes*, Faber and Faber, 1985.

The USSR against G. H. Hardy, page 66

A. M. Vershik, '"Nature" and Soviet Censorship', in *The London Mathematical Society Newsletter*, 218, July 1994, p. 12.

Poetry and Mathematics, page 68

Jonathan Holden, 'Poetry and Mathematics', in *Georgia Review*, vol. 39, 1985, p. 770.

If only . . ., page 69

Oysten Ore, *Niels Hendrik Abel, Mathematician Extraordinary*, Chelsea Publishing Company, New York, 1974, p. 230.

Galois, the schoolboy, page 69

Quoted in George Sarton, *The Life of Science*, Henry Schuman, New York, 1948, p. 87.

Mancala, a mathematical game, page 70

Claudia Zaslavsky, *Africa Counts: Number and Pattern in African Culture*, Lawrence Hill Books, New York, 1979, pp. 116, 124–5.

With a little help from my friends, page 72

J. W. McReynolds, 'George's Problem', in *Scripta Mathematica*, vol. 15, 2, June 1949.

Look inside the box, page 72

G. W. Leibniz, *Mathematische Schriften*, quoted in G. Polya, *Mathematical Discovery*, vol. 2, Wiley, New York, 1965, p. 99.

The value of experience, page 72

I. Todhunter, *The Conflict of Studies*, 1873, p. 17, quoted in *The Mathematical Gazette*, vol. 32, 300, 1948, p. 186.

The Dining Philosophers Problem, page 73

David Harel, *Algorithmics: the Spirit of Computing*, Addison-Wesley, 1987, p. 287.

Full Marx, page 74

Mathematical Digest, 22, January 1976, p. 7.

The basic facts, page 74

Entry, 'Dirichlet', in Clifton Fadiman (ed.), *The Faber Book of Anecdotes*, Faber and Faber, 1985.

The deaths of two Jewish mathematicians, page 75

Mark Kac, *Enigmas of Chance: an Autobiography*, University of California Press, 1987, pp. 46–7.

Max Pinl, 'In Memory of Ludwig Berwald', *Scripta Mathematica*, vol. 27, 3, 1964, pp. 195–6.

Hardy visits Ramanujan, page 77

C. P. Snow, *Varieties of Men*, Macmillan, 1968, pp. 32–3.

Hilbert's breakfast, page 77

J. Mehra (ed.), *The Physicist's Conception of Nature*, Kluwer, 1973, reprinted in T. Ferris (ed.), *The World Treasury of Physics, Astronomy and Mathematics*, Little, Brown and Co., 1991, p. 604.

Fabre, spiders, and geometry, page 78

Jean Henri Fabre, *The Life of the Spider*, trans. A. T. de Mattos, Hodder, 1913.

Immersion, page 79

B. Bollobas (ed.), *Littlewood's Miscellany*, CUP, 1986, (formerly, J. E. Littlewood, *A Mathematician's Miscellany*, Methuen, 1953).
Michael Atiyah, in 'An Interview with Michael Atiyah', in *The Mathematical Intelligencer*, vol. 6, 1, 1984, p. 17.

Beauty versus effectiveness, page 79

Henri Poincaré, *Science and Method*, Dover, n.d.
G. S. Rousseau, *Organic Form, the Life of an Idea*, Routledge and Kegan Paul, 1972, p. 50.

Geometry rules, OK, page 79

Frederick the Great, *Oeuvres*, quoted in A. L. Mackay, *The Harvest of a Quiet Eye – a Selection of Scientific Quotations*, The Institute of Physics, Bristol, 1977, p. 59.

Lovers, page 80

A. L. Mackay, *The Harvest of a Quiet Eye – a Selection of Scientific Quotations*, The Institute of Physics, Bristol, 1977, p. 58.

A mathematical bent, page 80

E. G. R. Taylor, *Mathematical Practitioners of Tudor and Stuart England*, CUP, 1954.

Odd Data, page 80

R. Houwink (ed.), *The Odd Book of Data*, Elsevier Publ. Co., 1965, pp. 10, 21, 44.

Dr Peter Borrows, 'Queries', the *Guardian*, 19 March 1993.
A. B. Pippard, *The Elements of Classical Thermodynamics*, CUP, 1966, p. 7.

Genius and Insanity, page 81

C. Lombroso, *The Man of Genius*, Walter Scott, 1891, p. 73.

The mind of a mnemonist, page 82

A. R. Luria, *The Mind of a Mnemonist*, Jonathan Cape, 1969, pp. 131–2.

The needs of the architect, page 82

Vitruvius, *The Ten Books of Architecture*, trans. M. H. Morgan, Dover, New York, 1960, pp. 5–8.

Wall wisdom, page 84

John Bibby (ed.), *Quotes, Damned Quotes . . .*, 2nd edn, 1986, John Bibby (Books), Edinburgh, p. 38.

How many computers?, page 84

David Singmaster, *Chronology of Computing, Version* 4, March 1994, South Bank University, pp. 3, 34.

Taking it nice and easy, page 84

B. Russell and A. N. Whitehead, *Principia Mathematica*, CUP, 1927, vol. 1, p. 379.

Infinity and the Sublime, page 85

Edmund Burke, *A Philosophical Enquiry into the Origin of Our Ideas of the Sublime and the Beautiful*, 1759, part 2, sec. 8, p. 129.
D. Hume, *An Enquiry into Human Understanding*, sec. xii, part ii, sec. 124.

The mathematician's need to communicate, page 86

John Playfair, 'Dissertation, exhibiting a general view of the progress of Mathematical and Physical Science', supplement to 4th, 5th and 6th editions of *Encyclopaedia Britannica*, 1816–24.

The generosity of algebra, page 86

D'Alembert, quoted in R. E. Moritz, *On Mathematics*, Dover, 1942, no. 1702.
Quoted in R. L. Wilder, *Evolution of Mathematical Concepts*, Open University Press, 1974, p. 14.

Mathematical pictures in physics, page 87

R. P. Feynman, 'The development of the space-time view of quantum electrodynamics', in *Les Prix Nobel*, Norstedt and Söner, Stockholm, 1966, pp. 179, 190.

The abstract becomes familiar, page 88

Alexander Macfarlane, *Lectures on Ten British Mathematicians of the Nineteenth Century*, Chapman & Hall, 1916, pp. 82–3.

Gödel becomes an American citizen, page 89

John Barrow, *Pi in the Sky*, Oxford, Clarendon Press, 1992, p. 118.

The Captain's Story, page 90

C. Davies, *Mathematical Science, its Logic and Utility*, Cassell, Peter and Galpin, n.d., pp. 144–5.

Admirable effects, page 91

Henry van Etten, *Mathematical Recreations, Or a Collection of Sundrie excellent Problemes out of ancient and modern Phylosophers, Both useful and Recreative*, London, 1633.

John Horton Conway, page 91

R. K. Guy, 'John Horton Conway', in D. J. Albers, G. L. Alexanderson (eds.), *Mathematical People, Profiles and Interviews*, Contemporary Books Inc., 1984, pp. 43–4.

The illustration is from D. G. Wells, *You Are A Mathematician*, Penguin, 1995, p. 155.

Newton's confidence, page 93

D. Bentley, *Memoirs of Sir Isaac Newton*, vol. 2, p. 407, quoted in R. L. Weber (ed.), *A Random Walk in Science*, The Institute of Physics, 1973, p. 187.

Einstein's tame mathematician, page 93

W. Rotherstein, *Men and Memories*, quoted in *The Mathematical Gazette*, vol. 19, 234, 1935, p. 205, note 1017.

Mathematics and the military, page 94

Henry Peacham, *The Compleat Gentleman*, 1622, and Robert Anderson, *Cut the Rigging: and Proposals for the Improvement of Great Artillery*, 1691, quoted in *Aubrey on Education*, ed. J. E. Stephens, Routledge and Kegan Paul, 1972, p. 167, notes 3, 4.

A. G. Stewart, *The Academic Gregories*, Oliphant Anderson & Ferrier, 1901, p. 23.

Dirac on poetry, page 95

I. A. Richards, 'The Writer and Semantics', in *Arena*, special issue, 24, October 1965, p. 93.

That's showing 'em, page 95

Entry 'Million' in John May, *Curious Facts*, Secker & Warburg, 1981.

The Moore method, page 95

P. R. Halmos, *I Want to Be a Mathematician*, Springer, 1985, pp. 255–9.

Benjamin Banneker, page 97

H. Eves, *Mathematical Circles Adieu*, Prindle, Weber and Schmidt, 1977, pp. 139–40.

Verse and worse, page 98

The first verse is traditional. An alternative third line is 'The Golden Rule is a stumbling stool.' The moral is always the same! The couplet is also traditional and subject to variation. The third verse is quoted in John Bowers, *Invitation to Mathematics*, Blackwell, 1988, p. 20.

The language of art, page 98

Margit Staber, *Georges Vantongerloo: Mathematics, Nature and Art*, The Studio, April 1974, pp. 181–4.

Down with Maths, page 99

G. Willans and Ronald Searle, *Down with Skool*, Max Parrish, 1953.

Light on his feet, page 101

L. J. Mordell, 'Reminiscences of an Octogenarian Mathematician', in *American Mathematical Monthly*, vol. 78, November 1971.

A question of definition, page 101

Quoted in *American Mathematical Monthly*, vol. 93, 5, May 1986.

Gauss on the Higher Arithmetic, page 101

Quoted in R. C. Laubenbacher and D. J. Pengelley, 'Gauss, Eisenstein, and the "Third" Proof of the Quadratic Reciprocity Theorem: *Ein kleines Schauspiel*', in *The Mathematical Intelligencer*, vol. 16, 2, 1994, pp. 67–8.

The education of Harish-Chandra, page 102

V. Kumar Murty, 'Ramanujan and Harish-Chandra', in *The Mathematical Intelligencer*, vol. 15, 2, 1993, pp. 33–9.

Mathematics from prison, page 102

André Weil, *The Apprenticeship of a Mathematician*, trans. J. Gage, Birkhäuser, 1992, quoted in a review by Lawrence Zalcman, *The Mathematical Intelligencer*, vol. 15, 4, 1993, p. 66.

The Romance of Mathematics, page 103

H. Eves, *Mathematical Circles Adieu*, Prindle, Weber and Schmidt, 1977, p. 27.

Derision by repeated subtraction, page 104

R. K. Guy, 'Conway's Prime Producing Machine', in *Mathematics Magazine*, vol. 56, 1, January 1983, p. 27.

Boy from the word go, page 104

D. J. Albers, G. L. Alexanderson and C. Reid (eds.), *More Mathematical People*, Harcourt, Brace, Jovanovich, 1990, pp. 222, 224.

Kicks, page 105

Hubert Phillips, in the introduction to his collection *Question Time*, Dent, 1937.

G. H. Hardy, *A Mathematician's Apology*, CUP, 1969, pp. 87–8.

Euler becomes blind, page 106

C. Truesdell, 'Leonard Euler, Supreme Geometer (1707–1783)', in H. E. Pagliaro (ed.), *Irrationalism in the Eighteenth Century*, Case Western Reserve University Press, 1972, pp. 84–5.

A prodigious childhood, page 107

S. J. Heims, John von Neumann and Norbert Wiener: *From Mathematics to the Technologies of Life and Death*, MIT Press, 1980, pp. 5–9.

Games and situations, page 109

Leibniz, letter to De Montmort, 29 July 1715, quoted in W. W. Rouse Ball, *Mathematical Recreations and Problems*, 2nd edn, Macmillan, 1892, epigraph to part 1.

Leibniz, letter to Huygens, 8 September 1679, quoted in M. J. Crowe, *A History of Vector Analysis*, University of Notre Dame, 1967, p. 3.

Travelling salesman ants, page 109

Charles Arthur, 'Smart ants solve travelling salesman problem', in *The New Scientist*, 4 June 1994, p. 6.

Paul Valéry and mathematics, page 111

Pierre Féline, 'Memories of Paul Valéry', in *Paul Valéry, Moi*, trans. M. and J. Mathews, The Collected Works of Paul Valéry, Routledge & Kegan Paul, 1975, pp. 8, 90, 359, 381.

Grace Chisholm Young (1868–1944), page 112

D. J. Albers, G. L. Alexanderson and C. Reid (eds.), *More Mathematical People*, Harcourt, Brace, Jovanovich, 1990, p. 299.

'A paradox, a paradox, a most ingenious paradox!', page 113

The first paradox is equivalent to the *batting paradox*. Suppose that in the first half of the cricket season Watson scores (for example) 252 in 4 innings and Smith scores 84 in 1 innings. In the second half of the season Watson scores 84 in 3 innings while Smith scores 252 in 7 innings. Then a quick calculation shows that Smith has the higher average in the first *and* second halves of the season, taken separately, yet it is Watson who has the higher average over the whole summer! (D. G. Wells, *You Are A Mathematician*, Penguin, 1995, pp. 304, 327.)

The last two examples are taken from: G. J. Székely, *Paradoxes in Probability Theory and Mathematical Statistics*, Reidel, 1986, pp. 55, 164.

The Unreasonable Effectiveness of Mathematics in the Natural Sciences, page 114

Eugene Wigner, 'The Unreasonable Effectiveness of Mathematics in the Natural Sciences', in *Communications in Pure and Applied Mathematics*, John Wiley, 1960, pp. 1–14.

'Mastermind' in Africa, page 115

Claudia Zaslavsky, *Africa Counts: Number and Pattern in African Culture*, Lawrence Hill Books, New York, 1979, pp. 116, 124.

A military approach, page 115

Paul Halmos, 'Think it gooder', in *The Mathematical Intelligencer*, vol. 4, 1, 1982, p. 21.

The pain of noncommunication, page 115

Alfred Adler, 'Mathematics and Creativity', *New Yorker* magazine, 1972, reprinted in T. Ferris (ed.), *The World Treasury of Physics, Astronomy and Mathematics*, Little, Brown and Co., 1991, pp. 435–46.

Gauss's brain, page 116

Stephen Jay Gould, *The Mismeasure of Man*, W. W. Norton, 1981, p. 93.

The Oulipo, page 117

Warren F. Motte Jr., *OULIPO: a Primer of Potential Literature*, University of Nebraska Press, p. 132. This is the only extended account in English of the work of the group. It includes biographies of group members and detailed bibliographies of their work.

A Primer in Humility, page 118

G. G. Joseph, *The Crest of the Peacock*, Penguin, 1991, pp. 167, 212, 179, 199, 270, 278–9, 297–8, 191, 289, 293, respectively, except for the original of the Gregory–Newton interpolation formula, taken from Li Yan and Du Shiran, *Chinese Mathematics: a Concise History*, trans. J. N. Crossley and A. W.-C. Lun, OUP, 1987, p. 161.

Cut it out, page 120

Ian Stewart, *The Problems of Mathematics*, OUP, 1987, p. 225.

Gauss and Poincaré: one step ahead of history, page 120

D. Goldfeld, 'Gauss's class number problem for imaginary quadratic fields', *Bulletin of the American Mathematical Society*, vol. 13, 3, July 1985, pp. 25–6.

Hadamard, quoted in B. Mandelbrot, *Fractals: Form, Chance and Dimension*, W. H. Freeman, San Francisco, 1977, p. 256.

Cardan's madness, page 120

The passages are from, respectively: J. Hadamard, *The Psychology of Invention in the Mathematical Field*, Dover, 1954, pp. 122–3, and, C. Lombroso, *The Man of Genius*, Walter Scott, 1891, pp. 74–5.

Really?, page 122

Bertrand Russell, *Mathematics and the Metaphysicians, in Mysticism and Logic*, Penguin Books, 1953, p. 75.

A unique diagram, page 123

J. Hadamard, *The Psychology of Invention in the Mathematical Field*, Dover, 1954, p. 111.

Minkowski and Smith, page 123

Constance Reid, *Hilbert*, George Allen and Unwin, 1970, pp. 11–12.

Alexander Macfarlane, *Lectures on Ten British Mathematicians of the Nineteenth Century*, Chapman & Hall, 1916, pp. 92–106.

The line from Montaigne translates: 'Nothing but the truth is beautiful, only the truth is worthy of love.'

An unhappy competition, page 125

Based on, Constance Reid, *Hilbert*, George Allen and Unwin, 1970, p. 19.

Speaking the lingo, page 125

Quoted in S. Körner, *The Philosophy of Mathematics*, Hutchinson, 1968, epigraph. (Translated from Hermann Weyl's *Raum, Zeit, Materie*, Berlin, 1923, para. 18.)

Dirac and women, page 125

G. Gamow, *Thirty Years that Shook Physics*, Heinemann, 1966, pp. 120–21. According to Gamow's footnote, when he questioned Mrs Dirac on this story, she said that Dirac had actually said, 'This is Wigner's sister, who is now my wife.'

Magic squares, page 126

The simplest magic square is a square array of nine cells in which the numbers 1 to 9 are placed in an essentially unique manner so that every row, every column and both main diagonals, sum to the same total (which is 15).

Larger magic squares have the same properties, that the sums of the rows and columns and main diagonals are all equal. Variants may demand that the minor diagonals, or other features, are also included.

Paul Carus in W. S. Andrews, *Magic Squares and Cubes*, Dover, New York, 1960.

Which are the most beautiful?, page 126

David Wells, 'Which is the most beautiful?', in *The Mathematical Intelligencer*, vol. 10, 4, 1988, p. 30, and David Wells, 'Are these the most beautiful?', in *The Mathematical Intelligencer*, vol. 12, 3, 1990, pp. 37–41.

G. H. Hardy, *Ramanujan*, 1927, quoted in G. E. Andrews, *The Theory of Partitions*, Addison-Wesley, 1976, p. 177.

The arrival of Indian numerals, page 128

G. G. Joseph, *The Crest of the Peacock*, Penguin, 1991, pp. 311–13.

Surprise, page 128

Jacques Ozanam, *Cursus Mathematicus*, 1712, vol. IV, preface.

How does mathematics exist?, page 129

E. W. Beth and J. Piaget, *Mathematical Epistemology and Psychology*, Reidel, 1966, pp. 111–12.

I. Kleiner, 'Rigor and Proof in Mathematics', in *Mathematics Magazine*, vol. 64, 5, December 1991, p. 305.

The cost of everything, page 129

J. E. Stephens (ed.), *Aubrey on Education*, Routledge and Kegan Paul, 1972, pp. 141–52.

Keith Thomas, *Numeracy in Early Modern England*, pp. 131–2, Transactions of the Royal Historical Society, 5th series, vol. 37, 1987.

A fashion for algebra, page 130

Jacques Ozanam, *Cursus Mathematicus*, London, 1712, vol. I, p. v.

An encomium, page 130

Jacques Ozanam, *Cursus Mathematicus*, 1712, vol. II, p. [a4].

Mathematical feeling, page 131

V. A. Krutetskii, *The Psychology of Mathematical Abilities in Schoolchildren*, Chicago University Press, 1976, p. 347.

Hungarians solving problems, page 131

Quoted in Reuben Hersh and Vera John-Steiner, 'A Visit to Hungarian Mathematics', in *The Mathematical Intelligencer*, vol. 15, 2, 1993, pp. 16–17.

Studying the teacher, page 132

John Gay and Michael Cole, *The New Mathematics and an Old Culture: a Study of Learning among the Kpelle of Liberia*, Hold, Rinehart and Winston, 1967, p. 34.

The joy of maths, page 133

William Wordsworth, preface to the 2nd edition of *Lyrical Ballads*.

Newton keeps Sylvester working, page 133

Alexander Macfarlane, *Lectures on Ten British Mathematicians of the Nineteenth Century*, Chapman & Hall, 1916, pp. 110–11.

Simplicity in mathematical physics, page 134

'C. N. Yang and Contemporary Mathematics, Interview with D. Z. Zhang', in *The Mathematical Intelligencer*, vol. 15, 4, 1993, p. 19.

Aubrey on Education, page 135

J. E. Stephens (ed.), *Aubrey on Education*, Routledge and Kegan Paul, 1972, p. 66.

O. L. Dick (ed.), *Aubrey's Brief Lives*, Secker and Warburg, 1949, pp. xc–xci.

Appearances, page 136

H. V. Morton, *In Search of Scotland*, p. 8, quoted in *The Mathematical Gazette*, vol. 16, 1932, p. 13, note 835.

The *Ladies' Diary*, or, *Woman's Almanac* (1704–1841), page 137

Teri Perl, 'The *Ladies' Diary* or *Woman's Almanac*, 1704–1841', *Historia Mathematica*, vol. 6, 1979, pp. 36–53.

David Wells, *The Penguin Book of Curious and Interesting Puzzles*, Penguin, 1992, pp. 39–40.

The stimulus of mathematics, page 137

Vivian Gornick, *Women in Science*, Simon and Schuster, New York, 1983, p. 65.

Hamilton's poetical inspiration, page 138

Alexander Macfarlane, *Lectures on Ten British Mathematicians of the Nineteenth Century*, Chapman & Hall, 1916, pp. 39–40.

Clifford's powers of visualization, page 139

H. Eves, *Mathematical Circles Squared*, Prindle, Weber and Schmidt, 1972, p. 81.

The real and the fancy, page 139

Stephen Jay Gould, *The Mismeasure of Man*, Penguin, 1981, p. 239.

Mathematics and the imagination, page 139

Voltaire, *A Philosophical Dictionary*, 1881, 3, 40.

G. Mittag-Leffler, quoted by Havelock Ellis, *The Dance of Life*, Constable, 1923, p. 128.

D. Hilbert, quoted in the *American Mathematical Monthly*, vol. 100, 4, 1993, p. 697.

Leonardo on flight, page 140

G. A. Tokaty, *A History and Philosophy of Fluid Mechanics*, Foulis, Henley-on-Thames, 1971, pp. 45–7.

Chasles is taken for a ride, page 141

H. Eves, *In Mathematical Circles*, Prindle, Weber and Schmidt, 1969, pp. 39–40.

Horner's Method, page 142

A. de Morgan, 'Remark on Horner's Method of Solving Equations', in T. S. Davies (ed.), *The Mathematician*, vol. 1, 2, 1845.

Problem-solving, page 143

R. van Gulik, *The Chinese Maze Murders*, and G. K. Chesterton, *The Scandal of Father Brown*, 'The Point of a Pin', quoted in G. J. Székely, *Paradoxes in Probability Theory and Mathematical Statistics*, Reidel, 1986, p. ix.

The end of mathematics, page 143

Hermann Weyl (1944) quoted in Morris Kline, *Mathematics: the Loss of Certainty*, OUP, 1980, p. 319.

Mathematics and Poetry, page 144

Traditionally ascribed to Weierstrass (1815–1897).

F. Pollock, *Clifford's Lectures and Essays*, vol. 1, New York, 1901, p. 1, quoted in R. E. Moritz, *On Mathematics*, Dover, 1942, entry no. 1121.

The Mathematical Gazette, vol. 14, 196, 1928–9, p. 234.

Projection, a novel device, page 144

T. S. Davies (ed.), *The Mathematician*, vol. 3, 5, 1849.

That beautiful hypothesis, page 145

E. T. Bell quoted in Clifton Fadiman (ed.), *The Faber Book of Anecdotes*, Faber and Faber, 1985, entry 'Laplace'.

Obsessions, page 146

J. W. N. Sullivan, *The Contemporary Mind: Some Modern Answers*, Humphrey Toulmin, London, 1934.

Formulas for primes, page 146

G. H. Hardy, 'A formula for the prime factors of any number', *Messenger of Mathematics*, vol. 35, 1905, p. 146.

F. E. Browder (ed.), 'Mathematical developments arising from Hilbert's problems', *American Mathematical Society*, 1976, p. 331.

Babbage swimming in circles, page 147

Charles Babbage, *Passages from the Life of a Philosopher*, Augustus M. Kelley, New York, 1969, p. 8 and pp. 206–7.

Putting two and two together, page 148

Clare Harman (ed.), *The Diaries of Sylvia Townsend Warner*, Chatto & Windus, 1994, quoted in *The Observer Review*, 19 June 1994, p. 18.

The Death of Abel, aged 27 years, page 149

Oystein Ore, *Niels Henrik Abel*, Chelsea Publishing Company, New York, 1974, pp. 219–25.

The language of Nature, page 151

Galileo, *Saggiatore*, Opere VI.

Cardan's death, page 151

Bayle's Dictionary, English translation, 1710, quoted in D. George, *A Book of Anecdotes*, Hulton Press, 1958, p. 291.

Madame du Châtelet, page 152

A. Maurois, 'Voltaire', Daily Express Publications, n.d., pp. 73–6.

Voltaire's verse translates: Without doubt you will be celebrated for the grand algebraic calculations in which your spirit is absorbed. I would like to try it myself, but Alas! A + D – B does not equal 'I love you'.

Mathematical madness, page 154

Whitehead quoted by S. J. Heims, *John von Neumann and Norbert Wiener*, MIT Press, 1981, p. 116.

B. Russell, 'The Study of Mathematics', *Philosophical Essays*, 1910, p. 73.

Checkmate, page 154

Chess: *Quotations from the Masters*, compiled H. Hunrold, Peter Pauper Press, New York, 1972.

Emanual Lasker (1868–1941) took his degree in mathematics and in 1902 obtained his doctorate for his work on abstract algebraic systems.

Henri Fabre observes himself, page 154

J. H. Fabre, *The Life of the Fly*, Hodder and Stoughton, 1913, pp. 293–301.

Fly me to the moon, page 156

Constance Reid, *Hilbert*, George Allen and Unwin, 1970, p. 92.

Johanna Gauss, page 157

W. K. Bühler, *Gauss: A Biographical Study*, Springer, 1981, p. 49.

The grandeur of Gauss, page 157

Alexander Macfarlane, *Lectures on Ten British Mathematicians of the Nineteenth Century*, Chapman & Hall, 1916, p. 96.

Galois' last letter, page 158

Galois, trans. L. Weisner, in D. E. Smith (ed.), *A Source Book in Mathematics*, pp. 278, 284–5.

The subjection of infinity, page 159

Voltaire, *Letters on the English*, The Harvard Classics, vol. 34, Collier, New York, 1910, p. 128.

The role of analogies, page 160

Quoted by W. W. Sawyer, in *A Path to Modern Mathematics*, Penguin, 1971, p. 84.

H. Poincaré, *Science and Method*, Thomas Nelson, 1908, p. 21. Cf. Feynman's emphasis on 'seeing apparently different phenomena as essentially the same'. (*Feynman Lectures*, vol. 1, p. 28.)

Quoted by F. Cajori in *History of Mathematics*, p. 345.

Quoted by W. H. Young, 'The Mathematical Method and its Limitations', *Atti del Congresso Internatzionale Dei Matematici*, Bologna, 1928: publ. N. Zanichelli, Bologna, 1929, vol. VI, pp. 203–14.

Ulam's working methods, page 160

Françoise Ulam, his wife, writing in, S. M. Ulam, *Science, Computers and People*, ed. M. C. Reynolds and G.-C. Rota, Birkhäuser, 1986, pp. xx–xxi.

The talent of youth, page 161

T. L. Heath, *Mathematics in Aristotle*, Clarendon Press, Oxford, 1949, p. 276.

An extraordinary mind, page 161

The 'substantial project' to which Ulam turned, with all the other scientists at the Los Alamos laboratories, was the creation of the hydrogen bomb.

G.-C. Rota, 'The Lost Café', in *Contention*, vol. 2, 2 Winter 1993, pp. 41–61. The Lost Café is a reference to the Scottish Café in Lwow, in Ulam's native Poland, where many famous mathematicians used to meet before the war.

S. M. Ulam, *Adventures of a Mathematician*, Charles Scribner, New York, 1976.

Education and mathematics compared, page 163

George Eliot, 'Thomas Carlyle', in *Selected Essays, Poems and Other Writings*, Penguin 1990, p. 343. (Mary Ann Evans, who wrote under the name George Eliot, attended public lectures in mathematics at Bedford College, London.)

Retrograde analysis, page 163

W. W. Sawyer, *Prelude to Mathematics*, Penguin, 1955, pp. 34–6.

Plato versus Greek ingenuity, page 166

Plutarch, quoted in J. L. Coolidge, *The Mathematics of Great Amateurs*, Dover, 1963, p. 12.

Pascal on geometry, page 167

B. Pascal, 'Reflections on Geometry and the Art of Persuading', in R. W. Gleason (ed.), *The Essential Pascal*, Mentor, 1966, pp. 297–327. Pascal's argument, omitted here, for the absence of definitions in mathematics is that the very simplest, most basic ideas involved are understood by everyone. Nevertheless, Pascal's argument has a strikingly modern tone.

From primary school to the stars, page 168

Sydney Smith, quoted in Clifton Fadiman, *Any Number Can Play*, Avon Book Division, 1957, p. 102.

The second quote is by Yuri Matijasevič, in 'My Collaboration with Julia Robinson', in *The Mathematical Intelligencer*, vol. 14, 4, 1992, p. 44.

Voltaire on Newton's achievement, page 168

Voltaire, *Letters on the English*, The Harvard Classics, vol. 34, Collier, New York, 1910, pp. 117–18.

Hardy's insurance, page 169

Traditional.

Hilbert and existence, page 169

Constance Reid, *Hilbert*, George Allen and Unwin, 1970, pp. 154–5.

The Egyptians teach their children, page 170

D. E. Smith, *Our Debt to Greece and Rome*, Cooper Square Publishers, New York, 1963, p. 158, quoting Jowett's translation of Plato's *The Laws*, V, 202.

Physics or mathematics?, page 170

Based on Mark Kac, *Enigmas of Change: An Autobiography*, University of California Press, 1987, pp. xxiii–xxiv.

A small mistake, page 171

Augustus de Morgan, *A Budget of Paradoxes*, Longman Green, 1872, p. 173.

Dürer's *Melencolia*, page 171

T. Lynch, 'The geometric body in Dürer's engraving *Melencolia 1*', in *Journal of the Warburg and Courtauld Institutes*, vol. 45, 1982, pp. 226–30.

Michael Atiyah talking, page 173

Roberto Minio and Michael Atiyah, 'An Interview with Michael Atiyah', in *The Mathematical Intelligencer*, vol. 6, 1, 1984, pp. 10–11.

Sir Christopher Wren and the stocking men, page 174

O. L. Dick (ed.), *Aubrey's Brief Lives*, Secker and Warburg, 1949, p. 191.

Love at first sight, page 174

O. L. Dick (ed.), *Aubrey's Brief Lives*, Secker and Warburg, 1949, p. 309.

P. A. Schlipp (ed.), *Albert Einstein, Philosopher-Scientist*, Evanston, 1951, p. 9.

B. Russell, *The Autobiography of Bertrand Russell, 1872–1914*, Allen & Unwin, 1967, p. 36.

All quoted in *Ben-Ami Scharfstein, The Philosophers, Their Lives and the Nature of Their Thought*, Blackwell, Oxford, 1980, pp. 14–15.

Laplace on the course of the universe, page 176

P.-S. Laplace, *Théorie analytique de probabilité, 1812–1820*, introduction, quoted in A. L. Mackay, *The Harvest of a Quiet Eye – A Selection of Scientific Quotations*, The Institute of Physics, Bristol, 1977, p. 92.

Ada Lovelace (1815–1853), page 176

The bibliographical information is taken from J. G. Crowther, *Scientific Types*, Barrie & Rockliff, 1968, p. 297.

The quotes are taken from, respectively, Ada Lovelace, in *General Menabrea's Sketch of the Analytical Engine, Invented by Charles Babbage*, trans. Ada Lovelace, October 1842; R. Taylor (ed.), *Scientific Memoirs*, III, 1843, p. 694, quoted in E. A. Bowles (ed.), *Computers in Humanistic Research*, Prentice-Hall, 1967, p. 179.

Eddington, page 178

A. S. Eddington, Tarner Lecture, 1938, reprinted in A. S. Eddington, *The Philosophy of Physical Science*, CUP, 1939. The number is $2^{256} \times 136$. The factor 136 is 137 − 1 where 137 is the 'fine structure constant' on which Eddington had almost mystical views.

The Age of the World, page 178

Patterns of Thought: The Hidden Meaning of the Great Pavement in Westminster Abbey, Richard Foster, Jonathan Cape, 1991, pp. 101–2.

The flea theme, page 178

C. W. Kimmins, *The Springs of Laughter*, Methuen, 1928, p. 172.
Jonathan Swift, *On Poetry, a Rhapsody*, 1733, lines 319–44.
Augustus de Morgan, *A Budget of Paradoxes*, Longman Green, 1872.
L. F. Richardson, *Weather Prediction by Numerical Process*, CUP, 1922, p. 66, quoted in B. B. Mandelbrot, *Fractals: Form, Chance, and Dimension*, W. H. Freeman, 1977, p. 269.

Hitting the target, page 180

R. D. Clarke, 'An application of the Poisson distribution', *Journal of the Institute of Actuaries*, vol. 72, 1946, p. 48. Quoted in D. O. Koehler, 'Mathematics and literature', *Mathematics Magazine*, vol. 55, 2, March 1982.

Euler's advice to a preacher, page 180

Arago, in the Chamber of Deputies, 23 March 1837, quoted in V. E. Johnson, *The Uses and Triumphs of Mathematics*, Griffith, Farran, Okeden and Welsh, 1889, pp. 107–10.

Newton, page 181

Lord Byron, *Don Juan*, 10, 11.
Notes & Queries, 27 January 1887, quoted in Daniel George, *An Eclectic ABC*, Barrie and Rockcliff, 1964, pp. 123–4.

Hardy's eccentricities, page 182

C. P. Snow, *Varieties of Men*, Macmillan, 1968, pp. 19, 37.

The End of Platonism?, page 183

E. T. Bell, *The Development of Mathematics*, McGraw-Hill, New York, 1940, p. 510.
One way to tackle the problem of the mode of existence of mathe-

matics and mathematical objects is to compare mathematics carefully with abstract games such as chess and go, which certainly are human creations. This approach is taken – very loosely and informally – in the author's, *You Are a Mathematician*, Penguin Books, 1995.

The pleasure of proof, page 184

Quoted in Clifton Fadiman, *Any Number Can Play*, Avon Book Division, p. 102.

Kepler on the snowflake, page 184

J. Kepler, *The Six-Cornered Snowflake*, Clarendon Press, Oxford, 1966, pp. 39, 41.

Ada Lovelace: how mathematicians think, page 186

Ada Lovelace, quoted in Betty Alexandra Toole (ed.), *Ada, the Enchantress of Numbers*, Strawberry Press, California, 1992. Also quoted in a review of that book in, appropriately, Newsletter 2 of the How Mathematicians Work Group.

The origins of geometry, page 186

Greek Mathematical Works, trans. Ivor Thomas, vol. 1, *Thales to Euclid*, Heinemann, 1980, p. 145.

'The *J*-type and the *S*-type among mathematicians', page 187

G. H. Hardy, 'The *J*-type and the *S*-type among mathematicians', in *Nature*, 1934, 7, 134, p. 250.

R. B. Fosdick, *Hitler and Mathematics, A Review for 1942*, The Rockefeller Foundation.

The oldest puzzle in the world, page 188

David Wells, *The Penguin Book of Curious and Interesting Puzzles*, Penguin 1992, pp. 3–4.

Did you know?, page 190

Sources: various, including *Mathematical Digest*, 22, January 1976, p. 7.

George Bidder, the Calculating Boy, on himself, page 190

E. F. Clark, *George Parker Bidder: The Calculating Boy*, KSL Publishing, 1983, pp. 3, 5.

Problema and theorema, page 191

Pappus, Collection iii, in *Greek Mathematical Works*, II, 'Aristarchus to Pappus of Alexandria', trans. Ivor Thomas, Heinemann, 1980, p. 567.

Quoted in OED, 'problem', sense 4, and 'theorem', sense 1a.

Charles Hutton, *A Philosophical and Mathematical Dictionary*, 1815, vol. 2, pp. 244, 504.

C. Godfrey and A. W. Siddons, *The Teaching of Elementary Mathematics*, CUP, 1946.

Turing, page 192

J. Boissonade, 'Long-range inhibition', in *Nature*, vol. 369, 19 May 1994, p. 188.

D. Singmaster, *Mathematical Monuments*, version of October 1993.

The natural history of differential equations, page 192

G. Temple, 'Linearization and delinearization', in J. A. Todd (ed.), *Proceedings of the International Congress of Mathematicians*, CUP, 1958, p. 233.

Holy Relic, page 193

Guinness Book of Records, Guinness Publishing, 1992, p. 69.

Vipers, logs and all that, page 193

G. J. S. Ross, *Eureka*, 22, October 1959, pp. 20–21.

Rabbi Solomon's problem, page 194

D. I. Golovensky, 'Maxima and Minima in Rabbinical literature', in *Scripta Mathematica*, vol. 1, p. 53.

Soroban versus Electric Calculator, page 195

The Magic Calculator, Japan Publications Trading Company, Tokyo, 1964.

'The Japanese Soroban', *Mathematical Digest*, 43, April 1981, p. 24.

See also: D. Nelson, G. G. Joseph and J. W. Williams, *Multicultural Mathematics*, OUP, 1993, p. 72.

The *Pons Asinorum*, page 196

J. M. F. Wright, *Self-examinations in Euclid*, Cambridge, 1829, p. 10.

J. H. Webb, *Mathematics on Vacation*, University of Capetown, 1987.

H. Gelernter, 'Realization of a geometry-theorem proving machine',

Proceedings of an International Conference on Information Processing, Paris, 1959, reprinted in E. A. Feigenbaum and Julian Feldman (eds.), *Computers and Thought*, McGraw-Hill, New York, 1963.

Gauss and Monsieur Leblanc, page 198

Margaret Alic, *Hypatia's Heritage*, The Women's Press, 1986, pp. 151–2.

The Taylor series remainder, page 198

John Barrow, 'It's All Platonic Pi in the Sky', *The Times Educational Supplement*, 11 May 1993.

Free market maths, page 199

P. Norridge, 'Free market maths', *Eureka*, 51, May 1992, p. 39.

Emmy Noether (1882–1935), page 200

Albert Einstein, *New York Times*, 4 May 1935, p. 12, col. 5, quoted by Steve Abbott, in *PLUS*, no. 25, Autumn 1993.
John Bowers, *Invitation to Mathematics*, Blackwell, 1988, p. 167.

Sylvester leaves the United States in a hurry, page 200

H. H. Bellot, *University College London 1826–1926*, 1929, p. 186.
Alexander Macfarlane, *Lectures on Ten British Mathematicians of the Nineteenth Century*, Chapman & Hall, 1916, p. 108.

Chess, mathematics, and the infinite, page 201

'Alpha of the Plough', *Pebbles on the Shore*, The Wayfarer's Library, J. M. Dent, 1935, pp. 160–61.

Paul Painlevé, politician, page 202

J. W. N. Sullivan, *Contemporary Mind, Some Modern Answers*, 1934, pp. 137–40, both quoted in *Scripta Mathematica*, vol. 13, 3, p. 273.

A fraction of the work, page 203

Quoted in David Tall, *Mathematicians Thinking about Students Thinking About Mathematics*, University of Warwick, Mathematics Education Research Centre, 1993, p. 7.

The power of Calculation, page 203

Shakuntala Devi, *Figuring: the Joy of Numbers*, André Deutsch, 1977, epigraph.

Fermat's Last Theorem, page 203

Stan Wagon, 'The Evidence, Fermat's Last Theorem', in *The Mathematical Intelligencer*, vol. 8, 1, 1986, pp. 59–61.

Harold Edwards, *Fermat's Last Theorem: A Genetic Introduction to Algebraic Number Theory*, Springer, New York, 1977.

Simon Welfare and John Fairley, *The Cabinet of Curiosities*, Weidenfeld and Nicholson, 1991.

According to the Bank of England, the exchange rate in 1908 was 20.66 German marks to the pound: hence, 100,000 marks = £4,840.

The remarkable Nicolas Bourbaki, page 206

I. M. Yaglom, *Mathematical Structures and Mathematical Modelling*, Gordon & Breach, 1986, pp. 62–3.

P. R. Halmos, 'Nicolas Bourbaki', in *Scientific American*, May 1957, pp. 77–9.

D. Guedj, 'Nicolas Bourbaki, Collective Mathematician', J. Gray trans., in *The Mathematical Intelligencer*, vol. 7, 2, 1985, p. 19.

A detective story, page 207

Paul Halmos, *I am a Mathematician*, Springer, 1985, p. 93.

The Fool and nothing, page 207

Shakespeare, *King Lear*, I: iv.

Hilbert on Mathematical Problems, page 207

David Hilbert, 'Mathematical Problems', trans. Mary Newson, *Bulletin of the American Mathematical Society*, vol. 8, 1902, pp. 437–8.

Newton on chaos and chance, page 209

Quoted in M. C. Battestin, *The Providence of Wit*, University Press of Virginia, 1989, p. 8.

Samuel Butler at Shrewsbury School, page 210

'Gleanings Far and Near', 291, in *The Mathematical Gazette*, vol. 12, 291, 1924–5.

The road to mathematics, page 210

Quoted in C. F. Linn, *The Golden Mean: Mathematics and the Fine Arts*, Doubleday, New York, 1974, p. 96.

Wiener's and von Neumann's methods of working, page 210

S. J. Heims, John von Neumann and Norbert Wiener, *From Mathematics to the Technologies of Life and Death*, MIT Press, 1980, pp. 123–7.

Seki Kowa (1642–1708), page 211

C. C. Gillispie (ed.), *Dictionary of Scientific Biography*, Charles Scribner's, 1980, entry 'Seki'.

A. Hirayama, K. Shimodaira and H. Hirose (eds.), *Seki Takakazu Zenshu*, Osaka Kyoiku Tosho, 1974.

Vitruvian proportion, page 212

Vitruvius, *The Ten Books of Architecture*, M. H. Morgan trans., Dover, New York, 1960, p. 73.

Sylvester finds a home from home, page 214

Alexander Macfarlane, *Lectures on Ten British Mathematicians of the Nineteenth Century*, Chapman & Hall, 1916, pp. 120–21.

Galileo recommended by his publisher, page 214

'The Publisher to the Reader', in Galileo Galilei, *Dialogues Concerning Two New Sciences*, trans. H. Crew and A. de Salvio, Dover, New York, 1914.

The Latin quotation translates: 'I only wish I could discover the truth as easily as I can expose falsehood.' It is taken from Cicero, *De Natura Deorum*, I, 91.

Poisson and the pendulum, page 215

Carl B. Boyer, *A History of Mathematics*, Princeton University Press, 1985, p. 569.

The quote is from W. W. Rouse Ball, *A Short Account of the History of Mathematics*, Macmillan, 1912, p. 434.

The Final Oral Exam, page 216

C. L. Fefferman and G. B. Folland, *The Mathematical Intelligencer*, vol. 14, 4, 1992, p. 31.

Euler as a marine engineer, page 217

C. Truesdell, 'Leonard Euler, Supreme Geometer (1707–1783)', in H. E. Pagliaro (ed.), *Irrationalism in the Eighteenth Century*, Case Western Reserve University Press, 1972, p. 51.

The discovery of quaternions, page 218

Hamilton, writing in 1865 to his son, quoted in M. J. Crow, *A History of Vector Analysis*, University of Notre Dame Press, 1967, p. 29.

The ocean of truth, page 219

Quoted in R. L. Weber (ed.), *A Random Walk in Science*, The Institute of Physics, 1973, p. 203.
G. G. Joseph, *The Crest of the Peacock*, Penguin, 1991, p. 268.

Erdös lends a hand, page 219

M. Kac, *Enigmas of Chance*, University of California Press, 1987, pp. 90–91.

Mathematicians, the users and makers of signs, page 220

S. M. Ulam, *Adventures of a Mathematician*, Scribner, 1976, p. 294.
Leibniz quoted in F. Cajori, *A History of Mathematical Notations*, Open Court, 1929, p. 185.

Pythagoras before Pythagoras, page 221

G. G. Joseph, *The Crest of the Peacock*, Penguin, 1991, p. 118.

Hilbert as a mathematical physicist, page 221

Quoted from Weyl's obituary notice for Hilbert in 'Hilbert's sixth problem', in F. E. Browder (ed.), *Mathematical Developments Arising from Hilbert's Problems*, American Mathematical Society, 1976, p. 157.

A South American abacus, page 222

Report by Father José de Acosta, a Spanish priest, who lived in Peru from 1571 to 1586, on the Inca's use of an abacus. G. G. Joseph, *The Crest of the Peacock*, Penguin, 1991, p. 40.

Gödel and Kafka, page 222

R. Rucker, *Infinity and the Mind*, Birkhäuser, 1982, p. 165.

Sylvester's enthusiasm, page 223

Alexander Macfarlane, *Lectures on Ten British Mathematicians of the Nineteenth Century*, Chapman & Hall, 1916, p. 116.

The numerical record, page 224

G. G. Joseph, *The Crest of the Peacock*, Penguin, 1991, p. 24. An illustration of this bone and further details may be found in

D. Nelson, G. G. Joseph, J. W. Williams, *Multicultural Mathematics*, OUP, 1993, p. 26.

A non-obvious conclusion, page 224

B. Pascal, 'On geometrical demonstration', in B. Pascal, *The Provincial Letters, Pensées, Scientific Treatises*, ed. R. M. Hutchins, *Encyclopaedia Britannica*, 1952, p. 443.

J. Perry, 'Report of Conference', *The Mathematical Gazette*, vol. 5, 77, January 1909.

Nature, art and microscopes, page 224

G. Grigson (ed.), *Before the Romantics*, The Salamander Press, 1984, p. 4.

Arnol'd on reading mathematics, page 225

V. I. Arnol'd, 'Interview with V. I. Arnol'd', in *The Mathematical Intelligencer*, vol. 9, 4, 1987, p. 30.

An almost-lost work, page 225

R. L. Wilder, *Mathematics as a Cultural System*, Pergamon, 1981, p. 110.

From a Kindergarten Teacher, page 226

Quoted from 'From a nineteenth-century kindergarten teacher's report', in K. Keeb-Lundberg, 'Kindergarten mathematics laboratory – nineteenth-century fashion', in *The Arithmetic Teacher*, May 1970.

Cardano wheedles the secret of the cubic out of Tartaglia, page 226

J. Fauvel and Jeremy Gray (eds.), *The History of Mathematics: a Reader*, Open University Press, 1987, pp. 254–55.

Hot stuff, page 228

Carl B. Boyer, *A History of Mathematics*, Princeton University Press, 1985, pp. 598–9.

The quotation is from H. Eves, *An Introduction to the History of Mathematics*, Holt, Rinehart and Winston, New York, 1976, p. 374.

Newton = 10 × Dryden, page 228

J. Cohen, *Homo Psychologicus*, George, Allen & Unwin, 1970, p. 52.

Dreaming a solution, page 229

J. Hadamard, *The Psychology of Invention in the Mathematical Field*, Dover, 1954, p. 7.

Ramanujan's flash, page 229

S. R. Ranganathan (ed.), *Ramanujan, the Man and the Mathematician*, Asia Publishing House, 1967, pp. 81–2, recording the reminiscences of P. C. Mahalanobis.

Winston Churchill's vision, page 230

Winston Churchill, quoted in J. H. Webb, *Mathematics on Vacation*, University of Capetown, 1987, pp. 32–3.

Poets and mathematicians, page 230

W. H. Auden, *The Dyer's hand and Other Essays*, Faber and Faber, 1963.

Turing's test – the imitation game, page 231

A. M. Turing, 'Computing Machinery and Intelligence', in E. A. Feigenbaum and J. Feldman (eds.), *Computers and Thought*, McGraw-Hill, 1963, pp. 11–12, 19.

The many proofs of Pythagoras, page 233

E. S. Loomis, *The Pythagorean Proposition*, National Council of Teachers of Mathematics, Washington, 1968 (originally published 1940).

The account of Leonardo's proof is taken from D. G. Wells, *The Penguin Dictionary of Curious and Interesting Geometry*, Penguin Books, 1991, p. 206.

Mathematics and Military Affairs, page 233

The Autobiography of Edward, Lord Herbert of Cherbury (1582–1648), ed. W. H. Dircks, Walter Scott, 1888, p. 32.

Two Anagramatists, page 234

D. Gjertsen, *The Classic of Science*, Lilian Barber Press, 1984, p. 164, note 16.

J. F. Scott, *A History of Mathematics*, Taylor & Francis, 1960, p. 155.

Flaubert teases, page 234

Edward Kasner and James Newman, *Mathematics and the Imagination*, Bell, 1970, p. 158.

'Seniora Wrangler', page 234

'Seniora Wrangler', in *Mathematical Digest*, 28, July 1977, p. 17.

African river-crossing problems, page 235

David Wells, *The Penguin Book of Curious and Interesting Puzzles*, Penguin, 1992, p. 82.

Marcia Ascher, 'A River-Crossing Problem in Cross-Cultural Perspective', in *Mathematics Magazine*, vol. 63, 1, February 1990, pp. 26–8.

The uses of a sextant, page 236

Francis Galton, *Hereditary Genius*, Macmillan, 1892, p. 20.

Advice from Paul Halmos, page 236

P. R. Halmos, *I Want to Be a Mathematician*, Springer, 1985, p. 216.

Mathematics in solitary, page 236

Roger Cooper, 'The inside story', in the *Spectator*, 13 April 1991.

All horses are the same colour, page 237

No reference. A more complicated version appears in R. L. Weber (compiler), E. Mendoza (ed.), *A Random Walk in Science*, Institute of Physics, 1973, p. 34.

The mathematicians as preacher, page 238

O. L. Dick (ed.), *Aubrey's Brief Lives*, Secker and Warburg, 1949, p. 116.

Advanced maths, page 238

G. Willans and Ronald Searle, *How to be Top*, Max Parrish, 1954, p. 44.

Giant numbers, page 239

G. G. Joseph, *The Crest of the Peacock*, Penguin, 1991, pp. 242, 251.

Harvard in 1802, page 239

P. Abelson, *The Seven Liberal Arts*, AMS Press, New York, 1906, pp. 107, 117.

F. Cajori, *A History of Elementary Mathematics with Hints on Methods of Teaching*, Macmillan, New York, 1890, p. 60.

J. M. F. Wright, *Self-examination in Euclid*, Cambridge, 1829, pp. 112, 173.

Book VI of Euclid's *Elements* includes the kind of material on similar triangles and circle properties that were studied by secondary school pupils before the New Maths invasion of the 1960s. The 'Rule of Three' taught the student to solve problems in proportion: given that 'A is to B as C is to D', and given the values of three of the quantities, to find the fourth. It was an essential tool for the merchant and other users of practical arithmetic. The later books of Euclid include Greek number theory and other topics which are beyond present GCSE standard, but easier than A-level.

Newton on poets, poets on Newton, page 240

Adapted from 'Gleanings Far and Near', 325, in *The Mathematical Gazette*, vol. 12, 173, 1924–5, p. 454, quoting *Spence's Anecdotes*.

Quoted in F. R. Moulton, *Introduction to Astronomy*, New York, 1906, quoted in R. E. Moritz, *On Mathematics*, Dover, 1958, p. 167.

Scripta Mathematica, vol. 14, 3, Notes and Queries, p. 290.

W. M. Priestley, 'Mathematics and Poetry: How Wide the Gap?', in *The Mathematical Intelligencer*, vol. 12, 1, 1990, p. 16.

William Wordsworth (1770–1850), *The Prelude*, III, lines 58–63.

David Singmaster, *Mathematical Monuments*, privately published, 1993, p. 19.

A canonical position, page 242

V. Sukhomlinsky, *On Education*, Progress Publishers, Moscow, 1977, p. 146.

John Aubrey on mathematics and small children, page 242

J. E. Stephens (ed.), *Aubrey on Education*, Routledge and Kegan Paul, 1972, pp. 98–100.

Sir Geoffrey Chaucer is Chaucer the poet (1342/3–1400).

Unearthly powers, page 243

Archimedes' claim was recorded by Pappus of Alexandria, *Collectio*, Book VII, prop. 11.

Thomas Paine, *The Rights of Man*, introduction to part 2.

The Lanchester model, page 243

F. W. Lanchester, *Aircraft in Warfare, the Dawn of the Fourth Arm*, Constable, 1916, part reprinted in J. R. Newman (ed.), *The World of Mathematics*, vol. 4, pp. 2138–57.

M. Braun, *Differential Equations and their Applications* (short version), Springer, 1978, pp. 291–9.

The Immortal Dinner, page 244

The author is the painter Benjamin Haydon. The quote is from *The Autobiography and Memoirs of Benjamin Robert Haydon (1786–1846)*, ed. Aldous Huxley, P. Davies, 1926, p. 317. The dinner took place in 1817.

A loss to mathematical science, page 245

O. L. Dick (ed.), *Aubrey's Brief Lives*, Secker and Warburg, 1949, p. 58.

Blaise Pascal meets Descartes, page 245

A. E. E. Mackenzie, *The Major Achievements of Science*, vol. 1, CUP, 1960, p. 60, quoting Mary Duclaux, *Portrait of Pascal*.

The lion's claw, page 246

H. W. Turnbull, *The Mathematical Discoveries of Newton*, Blackie, 1945, pp. 41–2.

An unusual god, page 246

Quoted in John Barrow, *Pi in the Sky*, Clarendon Press, Oxford, 1992.

John von Neumann, by his daughter, page 247

Marina v. N. Whitman, 'John von Neumann: a personal view', in: J. Glimm, J. Impagliazzo, & I. Singer (eds.), *The Legacy of John von Neumann*, American Mathematical Society, 1980, pp. 1–4.

Hilbert, page 248

N. J. Rose, *Mathematical Maxims and Minims*, Rome Press Inc., Raleigh, 1988, p. 35.

Rousseau on geometry, page 248

Quoted in the *Mathematical Gazette*, vol. 22, 249, 1938, p. 97.
R. L. Archer (ed.), *Rousseau on Education*, Edward Arnold, 1916, pp. 135–6.

Dirac on mathematical beauty, page 249

P. A. M. Dirac, 'The evolution of the physicist's picture of nature', in *Scientific American*, May 1963, vol. 208–5, pp. 45–53.

P. A. M. Dirac, in M. Goldsmith et al. (eds.), *Einstein: the First Hundred Years*, Pergamon Press, 1980, p. 44.

Mathematical memories, page 249

L. J. Mordell, 'Reflections of a mathematician', *Canadian Mathematical Congress*, McGill University, 1959, pp. 26–7.

A strange agreement, page 249

G. Sierksma, 'Johann Bernoulli (1667–1748): His Ten Turbulent Years in Groningen', *The Mathematical Intelligencer*, vol. 14, 4, 1992, p. 26.

Michael – the pocket Calculator, page 250

Giles Jackson, 'Michael – the pocket calculator', in *The Times*, 6 June 1994.

Dr Johnson invents bases, page 250

James Boswell, of Dr Samuel Johnson, quoted in John Barrow, *Pi in the Sky*, Clarendon Press, Oxford, 1992, p. 118.

Mathematics, art and science, page 251

Atte Selberg, 'Reflections around the Ramanujan Centenary', in *Collected Papers*, vol. 1, Springer, 1989, p. 697.

Eureka!, page 251

Vitruvius on Architecture, vol. 2, trans. F. Granger, The Loeb Classical Library, Heinemann, 1934, Book IX, preface.

Poincaré's memory, page 253

H. Eves, *Mathematical Circles Squared*, Prindle, Weber and Schmidt, 1972, p. 97.

The thirteen duels, page 253

H. Eves, *Return to Mathematical Circles*, Prindle, Weber and Schmidt, 1988, p. 35.

The mysterious source, page 253

Albert Einstein, *Ideas and Opinions*, trans. S. Bargmann, Alvin Redman, London, 1954.

From googol to googolplex, page 254

Edward Kasner and James Newman, *Mathematics and the Imagination*, Bell, 1970, pp. 20, 23.

Primo Carnera was heavyweight boxing champion of the world 1933–4.

Dynamic verse, page 255

L. Campbell and W. Garrett, *The Life of James Clerk Maxwell*, Macmillan, 1884, pp. 404–8, 419–20.

Hero's powerful magic, page 256

The Pneumatics of Hero of Alexandria, a facsimile of the 1851 Woodcroft Edition, introduction by Marie Boas Hall, Macdonald, 1971, pp. 68, 72.

Understanding, page 259

Heaviside is quoted in W. W. Sawyer, *Mathematician's Delight*, Penguin, 1967, p. 216.

Brains versus mathematics, page 259

C. A. Coulson, *The Spirit of Applied Mathematics*, Clarendon Press, Oxford, 1953.

The death of Archimedes, page 259

B. Perrin, (trans.), *Plutarch's Lives*, William Heinemann, 1917, vol. 5, p. 487.

Briggs meets Napier, page 260

Quoted in J. L. Coolidge, *The Mathematics of Great Amateurs*, Dover, 1963, pp. 77–8.

The End of Mathematics is Nigh, page 261

R. L. Wilder, *Mathematics as a Cultural System*, Pergamon, 1981, p. 67.

Keith Devlin, *Mathematics: the New Golden Age*, Penguin, 1988.

'Let no one ignorant of geometry enter here', page 261

D. E. Smith, *Our Debt to Greece and Rome*, Cooper Square Publishers, New York, 1963, p. 59.

Paul Erdös makes an offer, page 261

Richard Guy, *Unsolved Problems in Number Theory*, Springer, 1981, dedication.

Paul Erdös, 'My Scottish Book "Problems" ', in R. D. Mauldin (ed.), *The Scottish Book: Mathematics from the Scottish Café*, Birkhäuser, 1981, p. 39.

Hardy on proof, and Ramanujan, page 262

G. H. Hardy, *Ramanujan*, CUP, 1940, pp. 15, 19.

A complex question, page 263

M. Kac, *Enigmas of Chance*, University of California Press, 1987, p. 126.

Sign of the Times, page 264

Martin Whitfield, 'Unsung gay hero of a secret war is honoured with road sign', the *Independent*, 30 December 1993.

Bertrand Russell dreams his fate, page 265

G. H. Hardy, *A Mathematician's Apology*, CUP, 1969, p. 83.

A surfeit of eggs, page 265

Charles Babbage, *Passages from the Life of a Philosopher*, Augustus M. Kelley, New York, 1969, pp. 197–8.

Boarding House Geometry, page 266

Stephen Leacock, 'Boarding House Geometry' (1910), in *The Bodley Head Leacock*, ed. J. B. Priestley, The Bodley Head, 1957, pp. 26–7.

A man of figures, page 267

John Steinbeck, *The Moon is Down*, in *The Short Novels of John Steinbeck*, Heinemann, 1954, p. 304.

No flies on von Neumann, page 267

This story is now traditional. The 'trick' to which the poser referred is to argue that the fly is flying for the whole of the one hour before the cyclists meet, and so it flies a total of 15 miles. No further calculation is necessary.

Scaling the peaks, page 268

R. P. Boas, *American Mathematical Monthly*, vol. 93, 5, 1986, p. 498. The A.M.S. is the American Mathematical Society.

Index

READ MORE IN PENGUIN

In every corner of the world, on every subject under the sun, Penguin represents quality and variety – the very best in publishing today.

For complete information about books available from Penguin – including Puffins, Penguin Classics and Arkana – and how to order them, write to us at the appropriate address below. Please note that for copyright reasons the selection of books varies from country to country.

In the United Kingdom: Please write to *Dept. EP, Penguin Books Ltd, Bath Road, Harmondsworth, West Drayton, Middlesex UB7 ODA*

In the United States: Please write to *Consumer Sales, Penguin USA, P.O. Box 999, Dept. 17109, Bergenfield, New Jersey 07621-0120.* VISA and MasterCard holders call 1-800-253-6476 to order Penguin titles

In Canada: Please write to *Penguin Books Canada Ltd, 10 Alcorn Avenue, Suite 300, Toronto, Ontario M4V 3B2*

In Australia: Please write to *Penguin Books Australia Ltd, P.O. Box 257, Ringwood, Victoria 3134*

In New Zealand: Please write to *Penguin Books (NZ) Ltd, Private Bag 102902, North Shore Mail Centre, Auckland 10*

In India: Please write to *Penguin Books India Pvt Ltd, 706 Eros Apartments, 56 Nehru Place, New Delhi 110 019*

In the Netherlands: Please write to *Penguin Books Netherlands bv, Postbus 3507, NL-1001 AH Amsterdam*

In Germany: Please write to *Penguin Books Deutschland GmbH, Metzlerstrasse 26, 60594 Frankfurt am Main*

In Spain: Please write to *Penguin Books S. A., Bravo Murillo 19, 1° B, 28015 Madrid*

In Italy: Please write to *Penguin Italia s.r.l., Via Felice Casati 20, I–20124 Milano*

In France: Please write to *Penguin France S. A., 17 rue Lejeune, F–31000 Toulouse*

In Japan: Please write to *Penguin Books Japan, Ishikiribashi Building, 2–5–4, Suido, Bunkyo-ku, Tokyo 112*

In South Africa: Please write to *Longman Penguin Southern Africa (Pty) Ltd, Private Bag X08, Bertsham 2013*

BY THE SAME AUTHOR

The Penguin Dictionary of Curious and Interesting Numbers

Why was the number of Hardy's taxi significant? How many grains of sand would fill the universe? What is the connection between the Golden Ratio and sunflowers? From minus one and its square root to numbers so large that they boggle the imagination, all you ever wanted to know about numbers is here. There is even a comprehensive index for those annoying occasions when you remember the name but can't recall the number. Kaprekar numbers? Ah, yes, of course . . .

You Are a Mathematician

This entertaining and informative introduction to mathematics begins with the secrets of triangles and the dazzling patterns formed by even the simplest numbers. It takes readers on 'a journey from the Greek mathematicians to quantum theory', and concludes with a challenging game. Mathematics is an invaluable scientific tool, yet mathematical thinking is very like a game, relying on cunning tactics, deep strategy and brilliant combinations – this book is an ideal guide to its potential and pleasures.

BY THE SAME AUTHOR

The Penguin Dictionary of Curious and Interesting Geometry

What do the Apollonian gasket, Dandelin spheres, interlocking polyominoes, Poncelet's porism, Fermat points, Fatou dust, the Voderberg tessellation, the Euler line and the unilluminable room have in common?

They all appear among the hundreds of shapes, figures, objects, theorems, patterns and properties in this collection of geometrical gems. From the simple circle to fiendish fractals, from billiard balls bouncing round a cube to geometry with matchsticks, from Pythagoras to Penrose tilings to pursuit curves, they are all here, with a comprehensive index to lead you to that triangle thingumajig, you know, the one where all the points lie on a line . . .

The Penguin Book of Curious and Interesting Puzzles

Wherever there are human beings, setting and solving problems – sometimes after hours of head-scratching – have always been among their principal passions. The Arabs drew on the intricacies of Islamic inheritances, the Chinese investigated magic squares – this collection of logical and mathematical puzzles, none requiring specialist knowledge or more than pencil, paper and a few counters, brings together examples from the earliest times up to the inexhaustible riches of the present day. Whether they concern Prisoner's Dilemmas, fast-breeding rabbits, liars and truthtellers or Prince Rupert's cube, one thing is sure: endless entertainment is guaranteed.